ROUTLEDGE LIBRARY EDITIONS:
EVOLUTION

Volume 15

FUNCTIONAL AFFINITIES OF MAN, MONKEYS, AND APES

FUNCTIONAL AFFINITIES OF MAN, MONKEYS, AND APES

A Study of the Bearings of Physiology and Behaviour on the Taxonomy and Phylogeny of Lemurs, Monkeys, Apes, and Man

S. ZUCKERMAN

Routledge
Taylor & Francis Group
LONDON AND NEW YORK

First published in 1933 by Kegan Paul, Trench, Trubner & Co. Ltd.

This edition first published in 2020
by Routledge
2 Park Square, Milton Park, Abingdon, Oxon OX14 4RN

and by Routledge
52 Vanderbilt Avenue, New York, NY 10017

Routledge is an imprint of the Taylor & Francis Group, an informa business

British Library Cataloguing in Publication Data
A catalogue record for this book is available from the British Library

ISBN: 978-0-367-27938-7 (Set)
ISBN: 978-0-429-31628-9 (Set) (ebk)
ISBN: 978-0-367-26599-1 (Volume 15) (hbk)
ISBN: 978-0-429-29415-0 (Volume 15) (ebk)

Publisher's Note
The publisher has gone to great lengths to ensure the quality of this reprint but points out that some imperfections in the original copies may be apparent.

Disclaimer
The publisher has made every effort to trace copyright holders and would welcome correspondence from those they have been unable to trace.

1. CHIMPANZEE. Genus *Pan* [Photo. by F. W. Bond

2. ORANG UTAN. Genus *Pongo* [Photo. by F. W. Bond

[*Frontispiece*

FUNCTIONAL AFFINITIES OF MAN, MONKEYS, AND APES

A STUDY OF THE BEARINGS OF PHYSIOLOGY AND BEHAVIOUR ON THE TAXONOMY AND PHYLOGENY OF LEMURS, MONKEYS, APES, AND MAN

By

S. ZUCKERMAN

M.A., D.Sc., M.R.C.S.

Research Associate, Yale University

Lately Anatomist to the Zoological Society of London

and Demonstrator of Anatomy, University College, London

Author of "The Social Life of Monkeys and Apes"

LONDON

KEGAN PAUL, TRENCH, TRUBNER & CO. LTD.

BROADWAY HOUSE, 68–74 CARTER LANE, E.C.

1933

Printed in Great Britain by Butler & Tanner Ltd., Frome and London

TO

G. ELLIOT SMITH

CONTENTS

LIST OF ILLUSTRATIONS

FOREWORD

THIS book gives a taxonomic and phylogenetic survey of the findings of diverse experimental investigations of lemurs, monkeys, and apes. So far as I know, this information has not been considered collectively from this point of view before. Students of mammalian classification seldom pay much attention to experimental biology, and experimental biologists as a rule have little time for taxonomy and phylogeny. Indeed there is a tendency, which has grown with the success of experimental methods in biology, to decry interest in these fields of systematic zoology; one of the signs is that in biological circles to-day the question of evolution usually attracts less attention as an account of the supposed histories of different groups of living organisms than as a discussion of general principles arising out of such experimental studies as genetics. This tendency to lay aside phylogenetic problems has not, however, affected the problem of the relationships of the group of mammals to which man belongs. Interest in this question is still as strong as it was fifty years ago, when it was greatly and vicariously enhanced by the conflict between the doctrine of man's animal descent and certain conventional social opinions. To-day this doctrine can be considered from a much wider point of view than was then possible, largely because of growth in our knowledge of the dynamic characteristics of the subhuman Primates—the functional characteristics with which this book is concerned. Since the phyletic significance of this information is usually a by-product—the facts are generally sought for more practical reasons—it is interesting to consider some of the ways in which investigations of function differ, in their bearings on systematic mammalogy, from a study so closely linked with taxonomy as comparative morphology.

Monkeys and apes are chosen as experimental subjects largely because they are the only animals suitable for the investigation of certain diseases, for example poliomyelitis, or for the analysis of physiological mechanisms, such as the menstrual cycle. In these studies they are frequently regarded simply as necessary substitutes for rats, guinea-pigs, or other more easily handled laboratory mammals, and the investigator may be altogether indifferent to the taxonomic significance of his work—assuming that it has such significance. As a result, systematists are often left in ignorance of observations that may, from their point of view, be of considerable value, and taxonomy remains the poorer until the observations thus made slowly filter into the body of general knowledge. Sometimes, however, monkeys are chosen as experimental subjects in the belief that they are more akin to man than are rabbits and dogs, and in the hope that information gleaned from their use will have a close bearing on human problems. At times, too, different kinds of monkeys may be studied in the expectation that a comparative approach will provide a solution for the main problem of an investigation; or occasionally, simply because of interest in the comparison itself. The interest in the comparison of primate types revealed in physiological and medical studies usually proves to be essentially the same preoccupation with phyletic relationships as that which characterizes comparative morphology. If there is any difference between these diverse fields of primate study in their exploitation of this interest, it is probably only one of degree. A survey of the literature suggests that present-day morphologists, perhaps because of the greater age of their subject, are more cautious with regard to the phyletic bearings of their work than are students of function.

The difference between the practice of comparative morphology and that of comparative psychology is much more significant, although to some extent it is also one of degree. Comparative morphology systematizes knowledge about the structure of different animals, and provides a means for their classification. Its more dominant preoccupation is not with the variations displayed in different species of animals by some single structure which can be isolated from the total organism, but with the light these variations throw on the phyletic relationships of the animals themselves. This is not so in comparative psychology. Here the emphasis is usually reversed, and interest is focussed mainly upon variations in the isolate that is being considered and upon the evolution of that isolate—for example, upon the evolution of something called "mind," not upon the evolution of the animals whose behaviour tells of this mind. The history of the subject gives the clue to this difference, for, according to many authorities, the scientific development of the study of animal behaviour was urged as a reaction against the introspective method of psychology, and as a search for objectively definable psychological characteristics. Thus, behaviour, as such, remained and still remains the central interest of comparative psychology: and while the subject has always included within its scope questions of the evolution of what is called intelligence—through such stages as reflexes, tropisms, and instincts—until very recently it rarely concerned itself with the differentiation among different animals and groups of animals of such aspects or characters of behaviour as can be isolated and defined.

In the past few years several comparative studies, manifesting this newer interest, have appeared, and there are many signs that this field of investigation is rapidly growing. If these studies are to align themselves with other systematic researches on Primates, the characters they consider must be classified as thoroughly as possible within their own fields before any attempts are made to correlate them with facts derived from other primate studies. This obvious point has often been overlooked. Thus some writers on animal behaviour have assumed, apparently because morphologists have declared the apes to be man's nearest-of-kin, that chimpanzees necessarily excel monkeys in, for example, memory and the use of instruments. Actually, of course, such a conclusion could be justified only by experiment. As the study of taxonomy has amply shown in the past, there is no *a priori* reason why the distribution of characters of behaviour that are *arbitrarily* chosen for study should be in accordance with classification by morphological characters; indeed, it may often prove that such characters of behaviour have no taxonomic value at all. It is plain that the student of behaviour is likely to be misled if he pins too much faith—with but few observations—to any single view of primate relationships based on morphological investigation. Wood Jones has voiced the suspicion, in his book *Man's Place Among The Mammals* (1929), that the laboratory worker studies apes "under the belief that they represent Man in the making," and he warns the experimenter against believing "that he is investigating an incipient human stage—a stage through which Man has passed, and which may, therefore, be expected to throw light upon the development of human characteristics." He goes on to say that the findings of these experimenters are findings "derived from a specialized and senile end-product of a phylogenetic line." Whether or not Wood Jones's opinions on phylogeny are accepted, his statements emphasize the fact that the student of behaviour must be fully conversant with the bases of morphological views of the phylogeny of the Primates if he wishes to use them as points of departure for speculations in his own field of study. Unless this is realized, or comparative studies of primate behaviour are given wider scope than they formerly have been, such investigations cannot be expected to throw much light on the probable line of human descent, and many evolutionary generalizations within this field will probably prove to be false.

It is not only taxonomy that would benefit if the functional characters of the Primates were extensively studied from the comparative point of view, for there can be no doubt that the systematic zoological approach provides an excellent method for co-ordinating data in such a way as to suggest problems for further investigation. This method of evaluation reveals the gaps in our present knowledge of functional differentiation, and at the same time frequently helps to throw light on the nature of functional characters regarded from the points of view of physiology or behaviour.

What is obscure in a functional activity of one species, for example man, may be clarified by some peculiarity in the working of the homologous process in a related species, and problems whose existence had not previously been realized might be revealed by this method of approach. In so far as the following pages give a taxonomic view of the distribution of primate functional characters, they are also to a large extent a presentation of problems dealing with the differentiation of these characters. In its taxonomic bearings the book is not intended as a final statement about the inter-relationships of different Primates. My aim has been, rather, to emphasize seldom-used approaches to the question of primate phylogeny, to show how little they have been systematically tried, and to indicate what I regard as the proper place of functional investigations in the study of the classification and evolution of the Primates.

In discussing facts which demonstrate the use of systematic and taxonomic methods in physiology and in the study of behaviour, I am aware that I may have omitted some facts which could very well have been included, although, so far as I know, the field of data available for this purpose is not yet wider than I have actually indicated. I may also be charged with having chosen to consider only those characters which fit present views about the subdivision of the order Primates. This is partly true, but the same charge could be made against the morphological characters considered in taxonomy. Although tails are recognized characters, no one classifies Primates according to the length of these organs, since the resulting system would not be adequate to denote the distribution of a very large number of other characters, morphological and physiological, whose correlation with each other is indicated by a different system of classification. The functional characters which merit most consideration in the classification of the Primates are those which show some degree of correlation not only amongst themselves, but also with characters in other fields of primate studies. I have freely used the term "functional character" to refer to any character that is revealed by methods used in the investigation of the dynamic, rather than the static aspects of organisms. In any case the distinction between function and structure in taxonomic discussion is probably largely artificial, since all the processes of the body presumably have a physical basis; gross structure may therefore be regarded as the expression of the peculiarities of chemical processes.

The subject-matter of this book was originally prepared as a contribution to the discussion on "Primates and Early Man," which took place at the 1932 York meeting of the British Association for the Advancement of Science. Part of the same material also formed the basis of a paper which I read before the Royal Anthropological Institute, London, on November 23rd, 1932 (an abstract of this communication was published in *Man*, Vol. 33, pp. 13–16). In preparing the book I have been very greatly helped by the encouraging advice and stimulating criticism of Professor G. Elliot Smith, of Professor J. P. Hill, of Professor W. E. Le Gros Clark, of Professor H. A. Harris and of Professor C. Daryll Forde. I owe thanks also to the Prosectorial Committee of the Zoological Society, especially to Sir Peter Chalmers Mitchell, who for the past five years greeted all my ventures into primate studies in that favourable way which greatly stimulates further venturing. I am also grateful to those of my new and temporary colleagues at Yale University, particularly to Professor Fulton and to Professor Yerkes, who were kind enough to read the proof and offer suggestions.

S. ZUCKERMAN.

CHAPTER I

PECULIARITIES OF CLASSIFICATION IN THE ORDER PRIMATES

SCIENTIFIC thought becomes increasingly difficult the less its material is amenable to quantitative treatment and the more it is related to deeply rooted emotional attitudes. This is well recognized in questions of human origins, and students in this field usually realize that they must subdue what anthropocentric interest they naturally have in their subject if they wish their conclusions to have a sound and unbiassed foundation. Thus Darwin (1871), a prominent exponent of objectivity, writes that the descent of man should be discussed in the same way as that of any animal. What, for instance, are man's relations to other vertebrates? Does he vary in bodily structure and mental faculties? If he does, are his variations "governed by the same general laws, as in the case of other organisms?" Has he given rise to "varieties and sub-races, differing but slightly from each other, or to races differing so much that they must be classed as doubtful species?"

A complete answer to questions as broad as these demands extremely wide study, particularly of human genetics, and far wider, indeed, than Darwin himself probably realized. Until such studies have been made, views on human descent must continue to have a more modest, but nevertheless not altogether insufficient basis in the findings of palæontology, and in the comparison of man with other animals (particularly with the sub-human Primates) in regard to structure, embryology and growth, physiology, susceptibility to disease (including parasitic infections) and behaviour. Some of these fields of relationship are, unfortunately, not so widely known or explored as are others, current views on the phylogeny of the Primates being largely based on the conclusions of comparative anatomy and the relatively slight evidence provided by palæontology. The physiological evidence, with the exception of the blood reactions, is seldom considered. The same may be said about the facts relating to disease and behaviour.

These omissions in the discussion of the classification and phylogeny of the Primates are unnecessary. Partly they are the results of too narrow a specialization in primate studies, for natural interest in human structure, and the difficulties of observing living animals, have combined to lay far too heavy an emphasis, in phylogenetic discussion, on comparative morphology. Partly they are also due to the absence of any effective co-operation between the systematist on the one hand, and the physiologist, or biochemist, or helminthologist on the other. Although the professional systematist is trying to express animal relationships as correctly as possible, he is generally compelled to draw his conclusions from the evidence of morphology alone. Even should he possess the necessary training in other fields, it is unlikely that he would have the time to examine the journals of physiology, or biochemistry, or experimental medicine, in search of facts which, though primarily of interest to students in these fields from an entirely different point of view, may nevertheless have bearing on the taxonomic questions he is studying. As a result of this, the real meaning of taxonomy—the science of the laws of arrangement or classification—has in some quarters been narrowed to mean classification on the basis of a few arbitrary structural characters. If this restricted interpretation is disregarded, little reflection is needed in order to realize that the consideration of phylogeny can have a much firmer taxonomic foundation in the order Primates than in other groups of mammals where, owing to lack of knowledge, it is usually of necessity limited to palæontology and more superficial anatomy.

This fact helps to explain why, apart from anthropocentric interest, the systematics of the order of animals to which man belongs has so peculiar an appeal. It also suggests why it is so controversial a subject. The characters by which taxonomic and phylogenetic relationships are traced may or may not be conspicuously related to one another; but, whether they are correlated or not, the more they are multiplied, the more scope do they provide for argument, and the greater is the resulting confusion.

There are, however, other explanations of the endless scientific controversies about human descent. For example, the phylogeny of the Primates is mostly a preoccupation of anatomists with a medical training, whereas animal phylogeny as a general study is a problem of professional zoologists. Naturally, therefore, a profound difference in point of view often exists between the human anatomists and the zoologists who take up the studies of the phylogeny of man and the classification of the Primates. The zoologist is warned against constituting a species with few individuals and on a small number of characters. Moreover, his greater experience of animal forms gives him a better opportunity for becoming the competent systematist by whose judgment a species can be defined—to use a "definition" of species that seems to be greatly in vogue at the present time.* He realizes, at least in theory, that if one group of animals differs from another in the absence of certain characters and the greater expression of others common to both groups, the true value of the difference can be assessed only statistically. The anatomist, on the other hand, usually approaches the problem from a totally different angle, long acquaintance with the human form giving him a somewhat unwarranted confidence in the systematics of the *Hominidæ*. When the anthropometrist (see Morant, 1926 and 1930) tells him that skulls which he has regarded as distinct types are, from the point of view of measurement, a sample from a homogeneous population, he may refuse to accept such a view. He may affirm (see Keith, 1931) that in his craniological studies he is guided by anatomical characters whose value can be assessed, for purposes of race identification, by the unaided eye—apparently forgetting for the moment that measurement can only increase the power of observation, and that he continually uses it where he is able. As I have tried to show elsewhere (Zuckerman, 1933), the main result of this arbitrary "anatomical" treatment of the bony remains of archaic men is that to-day we accept an altogether irrational classification for the family *Hominidæ*.

But in his acclamation of this peculiar point of view, the anatomist may partly be labouring under the influence of a widely spread impulse to hail new observations in the field of primate studies as unparalleled and unique in their significance, irrespective of their

* A similar "definition" occurs in Darwin's *Origin of Species*.

exact relationship to pre-existing knowledge. It is an attitude sometimes demonstrated also by palæontologists, and one of its effects is chaos in the classification of another family of Primates, the *Pongidæ* (apes). The somewhat arbitrary systematic reference of fossil Primates may be exemplified by the following instance. At a recent meeting of the Anatomical Society of Great Britain and Ireland, Hopwood, who speaks with authority as a palæontologist, exhibited a maxilla, together with some teeth, found in a Central African Miocene deposit (believed to be Lower Miocene). He pointed out that the teeth are of the *Dryopithecus* pattern, and that they closely conform to the classical description, given by Gregory (1916), of the corresponding chimpanzee teeth. This he interpreted as indicating that the fossils represent an ancestral form of chimpanzee, but rather than refer them to the genus *Pan* (chimpanzee) or to the genus *Dryopithecus*, he suggested the creation of a new genus, *Proconsul* (see Hopwood, 1933).

Arbitrary though the creation of new genera and species may be, there can be little justification for this proposal. None can be found on the score of the age of the teeth, for there is no obvious reason why the same genus should not have persisted more or less unchanged through several geological epochs. An actual instance of this is provided by the Lamp shells of the Cambrian, which palæontologists are quite satisfied to call by the generic name (*Lingula*) that is used for existing shells of the same kind. Thus unless justification for the creation of new genera of apes or man is to be sought on the score of convenience of reference, a skull bearing the characters of *Homo sapiens* should be classified as *Homo sapiens*, even if it were recovered from an Oligocene deposit. If the teeth described by Hopwood are as much like those of a chimpanzee as he stated, there is no reason why they should not be included among the chimpanzees in the genus *Pan*. Moreover, precedent for more conservative treatment of the teeth of fossil apes is well provided by the genus *Dryopithecus*, which consists of more than six species, the teeth of which would seem to differ far more amongst themselves than the teeth of *Proconsul* differ from the corresponding teeth of the chimpanzee.* Fortunately for those who wish to see a less sensational classification and phylogeny of Primates than is usual, the danger of creating genera on the strength of single teeth is sometimes strikingly apparent, as the instance of *Hesperopithecus* shows, even though such procedure may occasionally be brilliantly justified, as in the case of *Sinanthropus*.

Proconsul is not a solitary exception to an otherwise well-balanced discipline of classification and phylogenetic study. Indeed, cases of a similar kind are so numerous as to force the conviction that primate material has a peculiar power of overweighting the conclusions of its students. This is unfortunate. General interest demands the constant re-telling of the tale of human evolution. In the circumstances this tale becomes either distorted or more dogmatic than the evidence warrants. Whether we like it or not, almost every fossil primate form, at least in the first excitement following its discovery, is given some special significance in the story of man's descent (e.g. the Taung's fossil). Moreover, the importance, whether it be real or unreal, attached to the question of man's evolution, prevents any separation of primate classification from

* See Gregory and Hellman, 1926.

primate phylogeny. The two are firmly knit together. It is essential, therefore, that the genera within a single family such as the *Pongidæ*, or the species within a single genus, should be at least approximately equivalent in their taxonomic differentiation, if the theoretical condition is to be fulfilled that the sub-groups of a given genus or family can be regarded as more closely related to each other by descent than they are to the members of other genera or families. Biassed speculation on the descent of the Primates only too easily prevents this, and a classification once made soon bears fruit. New students will accept without question a classification in which the relative difference between two genera of a family may actually be less than that between two species of a single genus, and will automatically draw distorted phylogenetic conclusions. If only to prevent this, a strong case could be made for the delimitation of taxonomic groups on a basis of a sliding scale of artificial units.

CHAPTER II

THE DYNAMIC BASIS OF CLASSIFICATION AND EVOLUTIONARY THEORY

THE explanation of the processes of variation and the scientific assessment of the concept of selection are the critical points on which the validity of theories of evolution ultimately depend. Neither variation nor selection can be adequately studied except by experiment and statistical analysis, and it is undoubtedly true that such studies as experimental genetics have vitalized the evolutionary hypothesis by providing it with a physiology. Their influence can be recognized in all biological fields in which the evolutionary concept has to be considered, and the effect they have had on such a subject as palæontology is one of the more remarkable features of twentieth-century biology. Even palæontologists who believe that new characters arise through the adaptation of old ones to changed conditions and function, realize that their views require further examination in the light of modern experimental genetics.

The experimental method can also find a field of application in taxonomy, and in phylogeny—the dynamic interpretation which the evolutionary concept allows to be made of the ideal classification or "natural system" provided by taxonomy. An ideal classification would be based upon a complete analysis of the resemblances and differences of all the "characters" that could be isolated from the organisms under consideration. And in this connection, characters, to quote Robson (1928), "include every structure and property of the animal or plant, whether they be organs, cytological structure, physiological activities, habits or ecological relationships." The obvious difficulties that prevent the realization of such an ideal classification also constitute the obstacles which serve to separate taxonomy from phylogeny. If, however, these fields of study are to be adequate, they must always tend towards each other, however far circumstances may keep them apart. A restriction of outlook in either the theory or the practice of taxonomy may easily lead to irrational conclusions. Thus in discussing the descent of man, which more widely is the problem of the phylogeny of the Primates, it may be misleading to refer, as Gregory does (1922), to "the taxonomic, palæontological and anatomical evidence", as though there existed a distinction between the taxonomic and anatomical evidence bearing on the problem, and as though the one kind of evidence supported the other. It is even more dangerous to base a classification on too few characters, or on characters weighted for arbitrary reasons. As Crow (1926) has pointed out, "each character, as it is observed in nature, must be accorded equal weight; there can be no evaluation of characters on the nature of the characters themselves." Such value ultimately depends upon the positions taxonomic characters assume in a scale showing their distributions within the group of organisms in which they are manifested. In practice, of course, organisms are classified according to purely superficial resemblances, but as Crow has written, it has been established in many cases that if a number of forms are arranged "in respect to the degree of development of one character they are also *ipso facto* arranged in a series in respect to other characters." In other words, characters are often correlated with one another, and we may, with justification, refer to the "character complex" of a species. Unfortunately the many cases which do not show such correlation of characters provide only too ready an explanation for the controversies of systematic biology.

A rational definition of "characters" such as that provided by Robson does away with any necessity to excuse the introduction of data relating to function into discussions of taxonomy or phylogeny.* If precedent for such a step be required, it is only necessary to remember the dynamic basis Cuvier gave to his "Règne Animal" (1817), or the more recent welcome given by systematists to Nuttall's classic researches (1904) on the serum precipitin reaction. Unfortunately, however, experimental biology is a young subject, and knowledge of the extent to which physiological differentiation occurs in the living world is very limited. It is unusual for enough facts about function to be available, as they are in the case of the Primates, to justify an attempt to trace their phyletic ramifications within the narrow boundaries of a single order of animals.

Apart from its use in providing taxonomic characters, the experimental method has other obvious and important functions to fulfil in taxonomy. Thus it should

* The question is discussed at length by Robson. Crow's (1926) view "that physiology, as ordinarily understood, cannot affect the findings of phylogeny" is based on the assumption that physiology is only concerned with "explaining the life processes" in terms of physics and chemistry. The contents of any physiological journal, however, show that a large part of its function is also to discover and describe these processes, which vary considerably from one living form to another. Like morphological characters they can therefore be used in taxonomy.

be used where possible to test the value of those single morphological characters for which special virtue is claimed in classification. A possible use, for example, is provided by the problem recently raised by Regan (1930). Regan has proposed a new and very unorthodox phyletic classification of the Primates, based, almost entirely, on the straight or wavy character of the enamel prisms of the teeth. He has not, however, provided any reason to explain why this character, which was described by Carter in 1922, should be regarded as of greater consequence than the correlated series of morphological features on whose distribution depends the orthodox view. Because of this omission it might quite reasonably be doubted whether his suggestion is sufficiently valuable to demand examination. An altogether different light would be thrown on the character, however, if his view were supported by an experimental study showing the factors which determine whether enamel prisms should have straight or wavy edges. It is of course possible that "straight" and "wavy", in regard to enamel prisms, are discontinuous characters, and that adequate statistical study will establish Carter's conclusions that they are differentiated as such among the different groups of Primates. If this were proved, enamel prisms would become a genetically distinct character of taxonomic value, and *ipso facto* would demand *equal* consideration with other such characters. As yet, however, it has not been shown whether or not their shape is affected by ontogenetic age, or by the mineral and vitamin content of the diet. Until this is done, there is no good reason for investing them with any special phyletic significance even though they suggest a novel classification of the Primates.

There are other individual characters used in taxonomy which could first be investigated experimentally, but only one more—coat colour—need be mentioned here. This character is constantly referred to in primate systematics, and if the ideal were possible, its physiology and genetics would doubtlessly be considered before it is used to differentiate species. If experimental knowledge of this kind were available, it would meet such difficulties as those raised by the classification of the Bornean leaf monkeys, *Pithecus cruciger*, *P. chrysomelas*, *P. rubicundus*, of which the first named, although at present regarded as a distinct species, may actually be, as Banks (1929) has pointed out on the basis of coat colour, a hybrid of the other two.

Some workers would extend this use of physiological data even further. With faith in the assumption that physiological characters, for example the protein specificity of blood serum, are more fundamental in their phyletic significance than those of morphology (in the language of phylogenetic discussion, that they are due to heritage rather than to habitus), they would relegate to them the function of a deciding issue in controversies occasioned by conflicting morphological taxonomic characters. There is no logical basis for this point of view. Function cannot be stressed more than form. Their separation from each other, and their individual isolation from the intact working organism, are both essential, but to a large extent necessarily incomplete, operations. The same is true of the abstraction of biochemical facts. Form, chemistry, and function are indissolubly united, and it may truly be said that from the taxonomic standpoint all the characters of the body, whatever their nature, have a fundamental equivalency.

Characters can be taxonomically differentiated amongst themselves only as they have a wide or narrow distribution among the organisms exhibiting them. To discuss them on the basis of *ex cathedra* judgments of their relationship to heritage or habitus is to attempt the impossible, and to ignore in greater part the lesson taught by the experimental study of genetics and growth.

Thus far I have been discussing functional characters in their bearing on taxonomy. On this level of discussion their application to the problem of phylogeny is the same as that of morphological taxonomic characters. As they may throw yet a different light on the problems of man's evolution, it is necessary to realize clearly that phylogeny as understood in systematic biology is an attempt to reconstruct, on the fundamental basis of the concept of evolution, the historical stages in the evolution of existing types. This it does by correlating a comprehensive taxonomy of existing and fossil animal forms with the spatial and temporal characters of these forms, the first being denoted by geographical distribution, and the second by the time sequence of the rocks. If animal classification is to some extent an arbitrary process, in so far as it demarcates groups which actually may have no natural definition, the phylogeny of systematic zoology is even less conclusive, and particularly so in its attempts to reconstruct the wider and the finer ramifications of the evolutionary process. Attempts to trace evolution in and out of a maze of genera and species are countered not only by the deficiencies of a somewhat arbitrary taxonomy, but also by the ever-present possibility that parallel and convergent evolution may have taken place among morphologically similar forms. The increasing number of students who accept the principle of orthogenesis (in the sense of a persistent evolutionary trend within a group of related organisms) is a striking testimony to the realization of this fact, and a warning against the acceptance, as scientifically valid, of the phylogenetic views of those students who are still only too ready to press their evolutionary studies too far on inadequate evidence.

Functional characters, however, help in revealing another, and possibly more interesting, side of the evolutionary process. It so happens that man's inclusion in the order Primates enables him to be used as a fixed point when the evolution of this group of animals is considered. Relative to other organisms, his most outstanding human characteristic is human intelligence, and the main emphasis in the story of man's evolution has always been laid on inferences, drawn from structural characters, concerning the functional stages through which man may have passed before his behaviour became what it is to-day. This story of change has in recent years become more than inference from anatomy, for a sound beginning has been made in the experimental investigation of the behaviour and cerebral function of sub-human Primates. It is still only a beginning, but the facts already discovered provide far more than corroboration to the inferences of anatomy, and, within the limitations of the comparative method, help one to visualize a dynamic human evolution. Here, perhaps, is a more valuable use for the study of physiology and behaviour than the mere provision of taxonomic characters.

In brief, the data derived from investigations of function may be used in studies of the taxonomy and phylogeny of the Primates in three ways; firstly, as simple taxonomic characters; secondly, in order to supplement our knowledge of morphological characters used in classification; and thirdly, as a basis for a "functional phylogeny". An attempt to use such data in this way is made in the following chapters, where they are discussed in relation to the accepted system of the classification of the Primates based upon morphological characters. In this discussion, as I have stated in the Foreword, the term "functional characters" is freely used to refer not only to characteristics of physiology and behaviour, but also to any characters that do not readily come under the heading of morphology.

CHAPTER III

THE SUBDIVISION OF THE ORDER PRIMATES

AMONG the more recent definitions of the criteria which a mammal must satisfy before it can be considered to be a Primate is that of St. George Mivart (1873*a*) Primates are

> "unguiculate claviculate placental mammals, with orbits encircled by bone; three kinds of teeth, at least at one time of life; brain always with a posterior lobe and calcarine fissure; the innermost digits of at least one pair of extremities opposable; hallux with a flat nail or none; a well-marked caecum; penis pendulous; testes scrotal; always two pectoral Mammæ."

These morphological characters are generally believed to represent a primitive mammalian condition, so that it may be truly said that the Primate, except for its general tendency to cerebral development, is relatively a non-specialized mammal.

A fairly orthodox classification of the animals which comply with St. George Mivart's criteria is given on p. 17, opposite. In slightly modified form, this list is chiefly a synthesis of the views of Pocock (1906, etc.) and of Schwarz (1926, etc.). The arrangement is based almost entirely on a consideration of external taxonomic characters, such as coat colour, relative proportions, and form of the genitalia. Nevertheless, in its main essentials the classification also satisfies the views of those students (e.g. Elliot Smith, Hill) who base their conclusions on more detailed anatomical

ORDER PRIMATES

Sub-Ord. LEMUROIDEA
(Grade Strepsirhini, Pocock)

- Series Lemuriformes (Madagascar)
  - Family Lemuridæ
  - Indrisidæ
  - Daubentoniidæ
- Series Lorisiformes (Africa and Asia)
  - Family Lorisidæ
  - Galagidæ

Sub-Ord. TARSIOIDEA
(Grade Haplorhini, Pocock)

- Family Tarsiidæ (Asia)

Sub-Ord. PITHECOIDEA
(Grade Haplorhini, Pocock)

- *Div. Platyrrhini* (South America)
  - Family Hapalidæ
  - Cebidæ
- *Div. Catarrhini* (Africa and Asia)
  - Family Cercopithecidæ
    - Sub-Fam. Cercopithecinæ
      - Gen. Mandrillus (Mandrill)
      - Papio (Baboon)
      - Theropithecus (Gelada baboon)
      - Cynopithecus (Black Ape of Celebes)
      - Macaca (Macaque)
      - Cercocebus (Mangabey)
      - Erythrocebus (Patas monkey)
      - {Cercopithecus (Guenon)
      - {Miopithecus (Talapoin)
    - Sub-Fam. Colobidæ
      - Gen. Presbytis (Langur)
      - Rhinopithecus (Snub-nosed langur)
      - Nasalis (Proboscis monkey)
      - Colobus (Guereza)
  - Family Hylobatidæ. Gibbon
  - Family Pongidæ. Gorilla
    - Chimpanzee
    - Orang
  - Family Hominidæ

study. It should be noted that the monkeys, apes, and man are referred to a sub-order called Pithecoidea, not Anthropoidea. Otherwise the nomenclature used in this book is mainly that of Flower (1929).

In order to facilitate reference in later chapters, the classification is given in some detail in the case of the catarrhine division of the sub-order Pithecoidea. Knowledge about systematic differentiation in the fields of physiology and behaviour is not yet available with respect to other groups.

The scheme given could have been altered without seriously influencing its phyletic implications by adopting certain well-recognized views. For example, the gibbons could have been denied familial rank and included with the *Pongidæ* (see Sonntag, 1924), or the sub-family *Colobidæ* could have been termed a separate family (see Pocock, 1925*c*).

Other variations have greater phyletic significance. Thus it was once usual to include the suborder Tarsioidea as a family of Lemuroidea (see Elliot, 1912; Gregory, 1916), but to-day this procedure is almost universally regarded as unjustifiable. Pocock (1918) is apparently alone in wishing to give subordinal rank to the family *Daubentoniidæ* of the series Lemuriformes. Some workers again (see Kaudern, 1910; Le Gros Clark, 1925) would include the Menotyphla within the primate order. This procedure would give definite expression to the accepted view that the roots of the primate stem are very closely intertwined with those of these primitive mammals. Wood Jones (1929) has added this view to his revival of the old idea (see, for example, Milne Edwards, 1871) that the Lemuroidea should be excluded from the order Primates, and suggests that a separate Tree-

[Photo. by F. W. Bond

3. BROWN LEMUR. Family *Lemuridae*

[Photo. by F. W. Bond

4. AYE-AYE. Family *Daubentoniidae*

[face page 18

shew-Lemur order should be created. There would be less opposition from orthodox circles to his acquiescence in the view of the relation of the Menotyphla to the lemurs than to his proposal to form a new order of mammals.

None of these modifications seriously affects the phylogenetic implications which are inherent in the commonly accepted classification of Primates. It is generally assumed that the Menotyphla of the upper Cretaceous gave rise to primitive prosimians, and that these divided into Lemuroidea and Tarsioidea. A group of the latter, some time before the close of the Eocene, evolved into primitive monkeys, which separated into a New World Platyrrhine group, and an Old World Catarrhine group from which the existing monkeys, apes and man developed. The nature of the relationship of the lemurs to the primitive Eocene tarsioids which are believed to have been the progenitors of the Pithecoidea is unknown, and some workers incline to the view that the Platyrrhine monkeys sprang from the *Notharctidæ*, a group of fossil Primates which Gregory refers to the series Lemuriformes of the sub-order Lemuroidea (for recent discussion see Straus, 1929). It is conceivable that one day a fossil may be found which will combine the characters which will satisfy Gregory's speculation (1922) that "the platyrrhine series . . . . started from small primates which had a tarsioid form of skull, but with certain marked differences in the front teeth and in the region of the auditory bullæ, which are more compatible with derivation from primitive members of the *Notharctidæ*." In the meantime it is useless to speculate that the prosimian ancestors of the Platyrrhini were widely different from those of the Catarrhini. Anatomical evidence (e.g. see Hill, 1932) gives support to the view that both divisions of the sub-order Pithecoidea were derived from a more primitive pithecoid stem, to which the platyrrhine monkeys approximate more closely than do the existing catarrhine Primates (Elliot Smith, 1930*b*). The two divisions may be regarded as diverging lines of descent from these primitive Pithecoidea.

In discussing variants of the orthodox classification of the Primates I have not referred to the "Tarsian hypothesis" of Wood Jones or to the diphyletic view of Regan, since both emphasize the phylogenetic rather than the purely taxonomic point of view. Wood Jones' opinion, as is well known, is that man evolved from tarsioid ancestors independently of other catarrhine Primates. Tate Regan's suggestion is that both the platyrrhine monkeys and the Tarsioidea evolved from Lorisiform, and the catarrhine Primates from Lemuriform, ancestors. This view he bases primarily on the character of the tooth enamel discussed in the preceding chapter (p. 11). Both of these unorthodox views are briefly considered in the concluding chapter of this book.

Since the history of the arrangement of the genera in the family *Cercopithecidæ* is not only of great interest, but also relevant to the argument of this book, it is reviewed here.

The arrangement given in the systematic table on page 17 is the one usual in modern works on the classification of the Primates. It is also quite common in pre-Darwinian literature, and so far as I can discover is founded on the memoir of G. Cuvier and E. Geoffroy Saint Hilaire on the Orang-Outangs.* In this work classification is chiefly based on degrees of elongation

* For a translation of this paper, see E. Griffith and others. 1827.

of the muzzle, and apart from the interpolation of the Sapajous (genus *Cebus*) between the apes and the genus *Cercopithecus*, it is very similar to the one in vogue to-day. Allowances must of course be made for the absence of forms described after 1800, and to errors due to imperfect and hearsay evidence.

In the *Tableau des Quadrumanes*, produced by E. Geoffroy Saint Hilaire in 1812, the Sapajous were removed from their anomalous position among the Old World Primates. Although other arrangements are given in many zoological treatises that appeared between 1800 and 1850, E. Geoffroy St. Hilaire's order of the genera *Cercopithecidæ* slowly became established as the accepted one. It appears, for example, in Kuhl (1820), in Temminck (1827), in Lesson (1827), and of course in his own and G. Cuvier's important *Histoire Naturelle des Mammifères* (1824). The genera *Cercopithecus*, *Cercocebus*, and *Macaca*, were often confused and telescoped in these earlier works, but they are clearly defined in Schinz's 1844 *Synopsis Mammalium* and in I. Geoffroy Saint Hilaire's 1851 *Catalogue des Primates*.

It is essential to realize that the pre-Darwinian order of the genera of the *Cercopithecidæ* was a purely systematic one. Its arbitrariness is amply revealed by the modifications introduced into it by different authors on the basis of such new taxonomic characters as the length of the tail, the presence of ischial callosities and buccal pouches, and the degree of elongation of the muzzle. Moreover the order was fixed in a period when the accepted zoological opinion was that species, separate and immutable, had been divinely created; a fact which makes it somewhat illogical to suppose that the pre-Darwinian classification was intended to convey any phyletic implications. It is true that the small band of dissenters from accepted zoological opinion included E. Geoffroy St. Hilaire, but so far as I can discover, there is no evidence that even he regarded the classification which he had helped to devise, as representing a phyletic scheme. His evolutionary leanings were not very manifest in the period in which he was interested in the taxonomy of the Primates (his well-known debates with Cuvier on the subject of the fixity of species date only from about 1830). In the circumstances one can but conclude that if the pre-Darwinian classification of Primates meant anything at all outside the narrow sphere of systematic zoology, it was accepted as revealing, not stages in the development of langurs from baboons, but the plan followed in the work of divine creation.

The idea of lower and higher in a phylogenetic sense could emerge only as part of the doctrine of descent, and in a system in which transformation plays a part. In this school of thought the post-Darwinian systematists were steeped, and when they inherited the pre-Darwinian classification of the Primates, they were not slow to interpret it phylogenetically. Thus the order of arrangement became an evolutionary one, so that to-day we may read such statements as "the Baboons are the lowest of the Catarrhine or Old World Monkeys" (Forbes, 1894), lowest being meant here in the evolutionary sense.* In spite of the frequency of such statements there does not appear to be the slightest evidence in favour of the view that baboons are actually lower in the evolutionary scale than are macaques, or that the genera of the *Cercopithecidæ* represent a linear

* It should perhaps be noted that such mention is not made in important works like Gray's 1870 *Catalogue of Monkeys*, and Flower and Lyddeker's 1891 *Introduction to the Study of Mammals*.

phylogenetic series. Those who have imposed this implication on the arrangement of the genera of the *Cercopithecidæ* have failed to realize either the arbitrary character of the classification they took over from pre-Darwinian naturalists, or the significance of the taxonomic evidence on which it is based.

CHAPTER IV

THE DIFFERENTIATION OF THE MECHANISMS OF REPRODUCTION

THE reproductive organs are specifically differentiated amongst nearly all orders of animals, and are of great diagnostic value in classification (see Robson, 1928). Reproductive mechanisms also are extremely varied, but they have been too little studied to be of use in the systematics of more than a very few groups, to which the Primates fortunately belong. In recent years research in this physiological field has made considerable advances, and has revealed many facts that bear on the inter-relationships of this group of animals.

For the purposes of the present discussion, the phenomena of reproduction may be considered under the following headings: duration of pregnancy; breeding season; reproductive cycle; and the changes in the sexual skin. I have included here, as well, data of great interest in themselves, relating to the duration of life in the sub-human Primates. This information is relevant to the subject-matter of this chapter, for although we have no knowledge of the duration of reproductive life in these animals, or of its relation to total life, we may reasonably suppose on the basis of our knowledge of human beings that there is some such relation.

[Photo. by F. W. Bond

5. SLENDER LORIS. Family *Lorisidae*

[Photo. by F. W. Bond

6. DEMIDOFF'S GALAGO. Family *Galagidae*

[*face page* 24

(*a*) The Duration of Life in the Sub-Human Primates

In considering this question first, my idea is to dispose of it before dealing with those problems which are more closely related to each other and to the question of reproduction itself.

As all who have studied the comparative longevity of animals have pointed out (e.g. see Lankester, 1870; Mitchell, 1911), conclusions on this subject can be based only on the "vague materials" constituted by the records of captive mammals. Because of this, it has been necessary to define with care the significance that can be attached to such figures as are relevant to the problem. Lankester's exposition of the various interpretations which in this connection can be made of the term "longevity", has been adhered to by most later writers, and the following of his points—all of which are fairly obvious—may be noted here.

Longevity varies among the individuals of a species, and the average longevity of its individuals gives the longevity of a species. Average or specific longevity, however, is not the same as potential longevity, which is the average period to which animals might live provided that they were not subject to ills, accidents, and struggles. Thus in some species vast numbers of individuals are destroyed in early phases of development. These catastrophes obviously affect only our ideas about the average longevity of the species, and not our ideas about the potential longevity of the survivors. Since it is almost impossible to discover the limits to the "potential longevity" of animals, we may for practical purposes assume it to be the greatest individual longevity of which we know.

A lengthy account of the duration of life in mammals, a large part of which is devoted to the Primates, has recently been published by Flower (1931). So far as I am aware, this report, which is based entirely on actual records, is the most reliable monograph on the subject ever published. I have used it freely in the following paragraphs.

Even when records relating to man are excluded, it is found that the Primates, for their size, form the longest-lived order of mammals. In the lemurs, the sexes seem to enjoy an equal longevity, but among monkeys, as in man, females seem on the average to live longer than do males. Some species, for example members of the genus *Erythrocebus*, almost always fail to live for any length of time in captivity. Others thrive exceptionally well. I have tried to summarize Flower's further data below, in table I. It must, of course, be remembered that his data are to some extent selected, in so far as he has purposely not considered animals dying for obvious pathological reasons within a short time of their arrival in a Zoological Garden. I, in turn, have selected Flower's data in drawing up the longevity table, not only for the sake of brevity, but also because Flower often refers only to the outstanding successes that have been achieved with different Primates in captivity, and not to average figures.

To what extent either the few "average lives" that it has been possible to include in table I, or the "longevity records", reflect "specific" and "potential" longevities it is impossible to say. Obviously, both columns give lower values than the true ones—since the unknown periods animals have lived before coming into captivity cannot be taken into account in the calculations. Again, while in some species (for

TABLE I

Family (or Genus).	Number of Genera (or Species).	Number of Individual Records.	Average Length of Life in Captivity (Years).	Maximum Longevity in Captivity (Years).
Gorilla gorilla: Gorilla	1	1	—	$7\frac{1}{12}$
Pan satyrus: Chimpanzee	1	20	11	$26\frac{2}{12}$
Pongo pygmæus: Orang	1	—	circa 8	$26\frac{6}{12}$
Hylobatidæ: Gibbons	2	—	circa 9	24
Cercopithecidæ (O.W. monkeys)			circa 7	46
Presbytis: Langur	—	—	—	15
Cercopithecus: Guenon	6	11	18	24
Erythrocebus: Patas	2	3	16	$19\frac{1}{12}$
Cercocebus: Mangabey	1	1	—	14
Macaca: Macaque	3	—	15	29
Cynopithecus: Black Ape	—	—	—	15
Papio: Baboon	—	—	circa 15	45
Mandrillus: Mandrill	—	—	circa 10	46
Cebidæ (N.W. monkeys):				
Lagothrix humboldtii: Woolly monkey	—	1	—	$5\frac{11}{12}$
Ateles ater: Spider monkey	—	—	—	7
Cebus: Capuchin	—	—	circa 10	25
Hapalidæ: Marmosets	—	—	—	16
Lemuridæ: Lemurs	4	119	$10\frac{6}{12}$	26
Lorisidæ: Lorises	—	—	—	10
Galagidæ: Galagos	—	—	—	$9\frac{9}{12}$
Daubentoniidæ: Aye-Ayes	1	4	$7\frac{2}{12}$	$8\frac{10}{12}$

example, baboons) the figures may be an accurate reflection of the real facts, in others (for example, gorillas) it is plain that the conditions of Zoological Gardens, though they be as good as can be achieved in present circumstances, are not good enough to allow the animals to thrive even reasonably well. In spite of the limitations of the data, comparison with the records for other orders of mammals shows, as I have already noted, that the Primates are among the longest lived of mammals.

(*b*) The Duration of Pregnancy

The length of gestation in lemurs is imperfectly known. E. Geoffroy and Cuvier (1824) estimated it as 111 days in the species *L. albifrons*, but according to Schmidt (1882), whose observations were made on a pair of animals belonging to the species *L. macaco* (= *L. niger*), it lasts about 145 days.

Nothing is known about the duration of pregnancy in *Tarsius*.

In the marmosets (family *Hapalidæ*, div. *Platyrrhini*, sub-order Pithecoidea) it lasts about 150 days. There are no reliable records regarding the family *Cebidæ* of the same division. In catarrhine monkeys (Rhesus macaques, common macaques, pig-tailed macaques, baboons) gestation usually lasts between 150 and 180 days.*

A reliable estimate of the length of gestation in a chimpanzee has recently been published by Tinklepaugh (1932*a*), pregnancy in the animal he observed having continued for nine lunar months. Information about the gestation of an orang-utan has also recently been put on record, the investigator, Aulmann (1932) noting that it lasted 275 days, or nearly ten lunar months. In man pregnancy continues for 280 days.

In so far as it is possible to generalize on such scanty data, it can thus be said that the duration of gestation in apes approximates more nearly to that of man than does the gestation period of monkeys.

(*c*) The Breeding Season

Pithecoidea.

Reliable data on this subject are few, but when critically analysed, they point to the conclusion that

* The sources of this information are given in full in Zuckerman, 1930 & 1932*a*.

all Old World monkeys about which accurate information is available breed at any time, although there may be seasonal variations in the birth rate (see Fig. 1). There is no reason to suppose that the apes are any exception to this generalization, while the only acceptable data about New World monkeys, both *Cebidæ* (*Ateles*, *Cebus* and *Alouatta*) and *Hapalidæ* (*Hapale*), show that they too breed at all times (for discussion and bibliography, see Zuckerman, 1931, 1932*a* & *d*).

Tarsioidea.

The Hubrecht data regarding *Tarsius spectrum* show that this animal breeds continuously, and that no significant seasonal variation in its birth-rate occurs (see Fig. 2). This statement is based on the analysis of almost 1,000 records (see Zuckerman, 1932*d*, for details).

Lemuroidea.

The members of this sub-order are geographically divided into the three groups, Mascarene, African, and Asiatic, the first group comprising the systematic division Lemuriformes, and the last two groups together forming the division Lorisiformes.

Available data about the breeding habits of the Lemuriformes are restricted to the genus *Lemur*. It appears to be fairly certain that members of this genus have a demarcated breeding season. As Figure 3 shows, the composite curve for 66 births in all except one species of this genus is almost identical with the frequency polygon for 29 births in the single species *L. fulvus*.

The breeding habit of the African members of the series Lorisiformes appears to be different from that

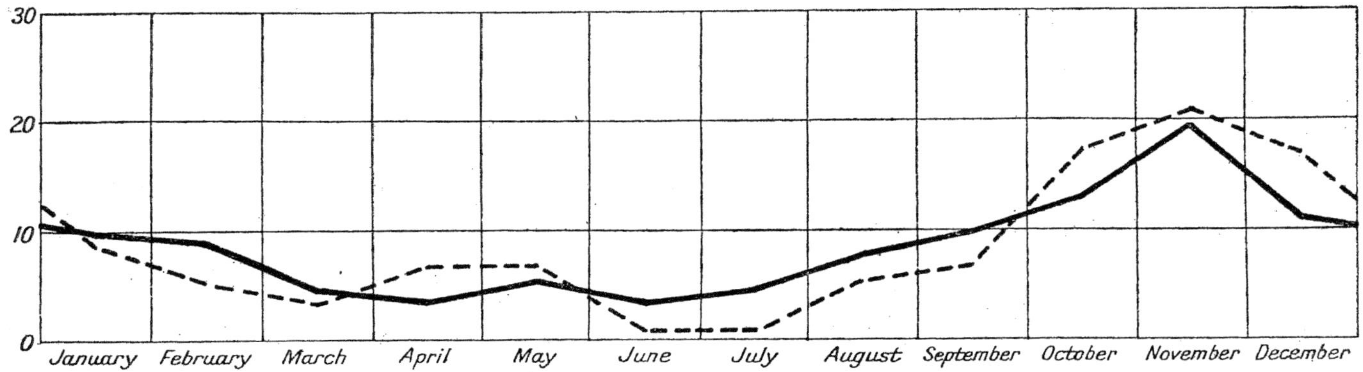

FIG. 1.—*PAPIO HAMADRYAS* AND *MACACA MULATTA*

Frequency polygons showing the monthly distribution, in percentages, of 87 Hamadryas baboon births that occurred in the London, Munich and Giza Zoological Gardens, and of the dates of conception of 118 Rhesus macaques that were born in various Gardens in Europe and North America. Data about macaques from Hartman, 1931. Data about baboons from Zuckerman, 1932*d*.

——— Baboons ------ Macaques

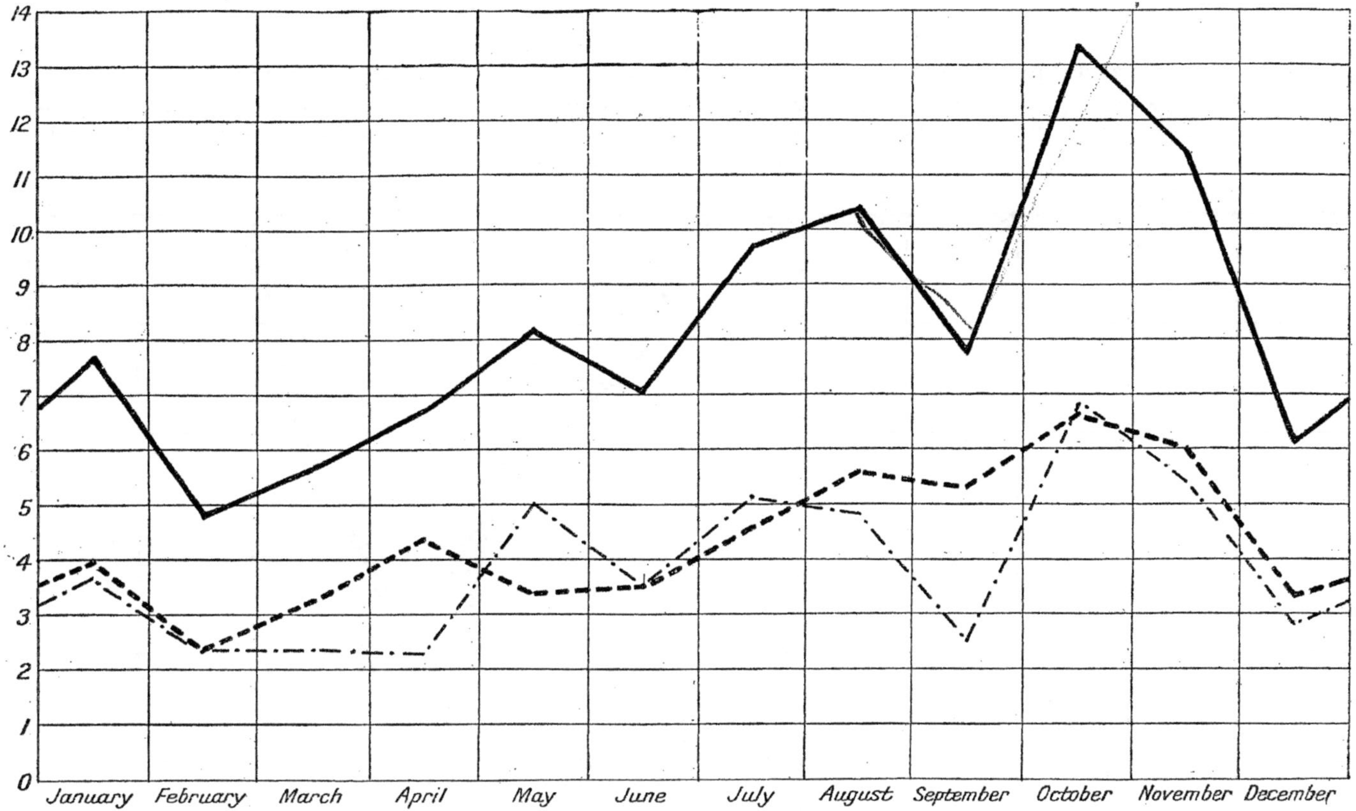

FIG. 2.—*TARSIUS SPECTRUM*

Frequency polygons showing the monthly distribution (*a*) of all specimens in the Hubrecht Collection, Utrecht, (*b*) of pregnant specimens, and (*c*) of non-pregnant specimens. Points plotted are percentages of the total number of all uteri in the collection (882). If the data are regarded as being adequate in number, the varying divergence from each other of the lines for pregnant and non-pregnant specimens may be taken to indicate seasonal fluctuations in birth-rate.

——— All specimens   - - - - - - Pregnant specimens   -·-·-·- Non-pregnant specimens

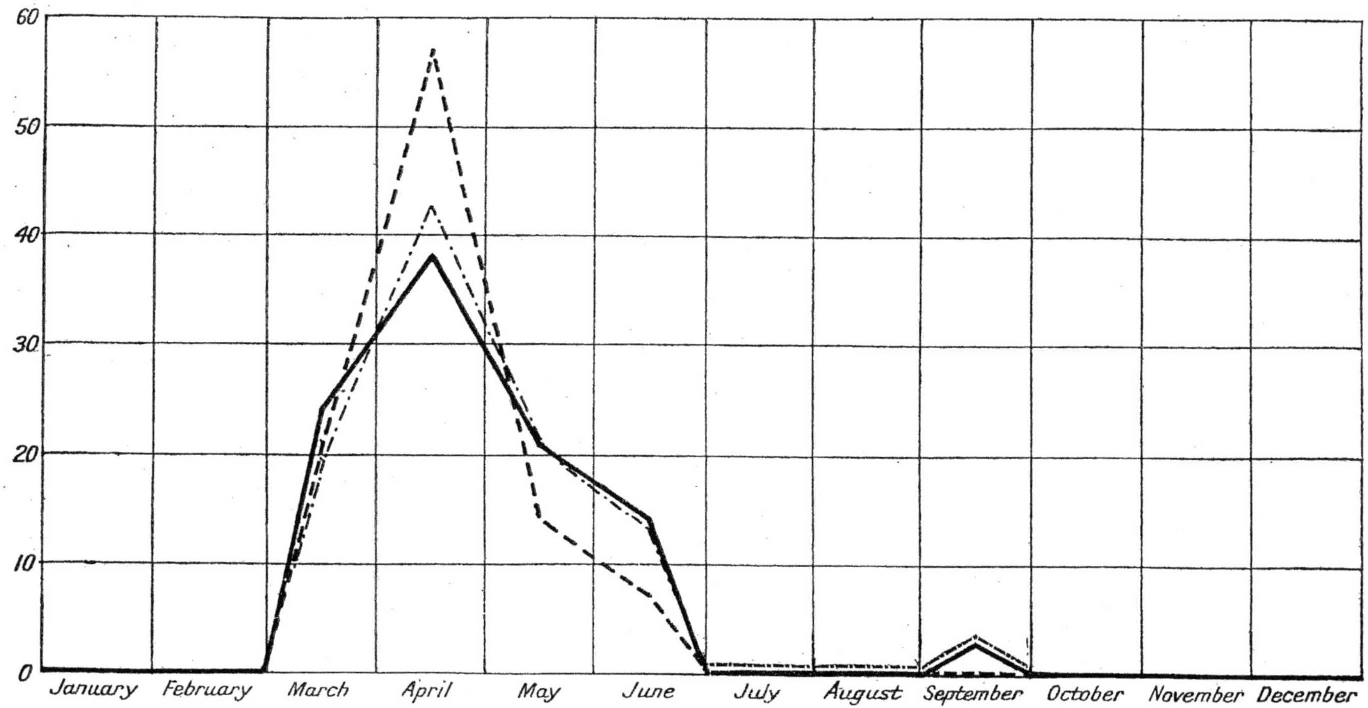

FIG. 3.—GENUS *LEMUR*

Frequency polygons showing the percentage monthly distribution of births of (*a*) Brown Lemurs (*L. fulvus*) 29, (*b*) Black Lemurs (*L. macaco*) 14, and (*c*) all lemurs *fulvus, macaco, rubriventer, mongoz*, 66. All births occurred in the London Zoological Gardens.

——— *L. fulvus* ------ *L. macaco* -·-·-·- All lemurs

of the Asiatic species belonging to the same subdivision of the Lemuroidea. Thus, the few available records regarding births in the genus *Galago* suggest that the animals belonging to this genus may have a restricted season, whereas the Hubrecht data for the slow loris (*Nycticebus coucang*) prove that this animal breeds throughout the year (see Fig. 4 and Zuckerman, 1932*d*).

The distinction in breeding habit between members of the Lemuroidea, on the one hand, and of the Tarsioidea and Pithecoidea on the other, not only seems to have value from the point of view of classification, but also appears to have phyletic significance.

The extent of an animal's breeding activity is affected by many external factors (e.g. food and warmth), and some writers, on the basis of experiments on starlings, ferrets and field mice, are inclined to generalize that an animal's breeding potentialities vary directly with the amount of light to which it is exposed. There can be no doubt that this is an unsound and unjustifiable generalization, for a critical survey of the evidence points clearly to the conclusion that factors inherent in themselves, and not immediate external conditions, determine whether mammals have *absolutely* restricted breeding seasons, or whether they are capable of breeding throughout the year. Thus, the same conditions of light do not stimulate the baboons and lemurs in the London Zoological Gardens to breed at the same time of the year. Fundamentally, an animal's breeding habit is part of its inherited physiological constitution.

In mammalian classification, the occurrence or non-occurrence of a breeding season has different taxonomic value in different groups. Thus the Axis and Red

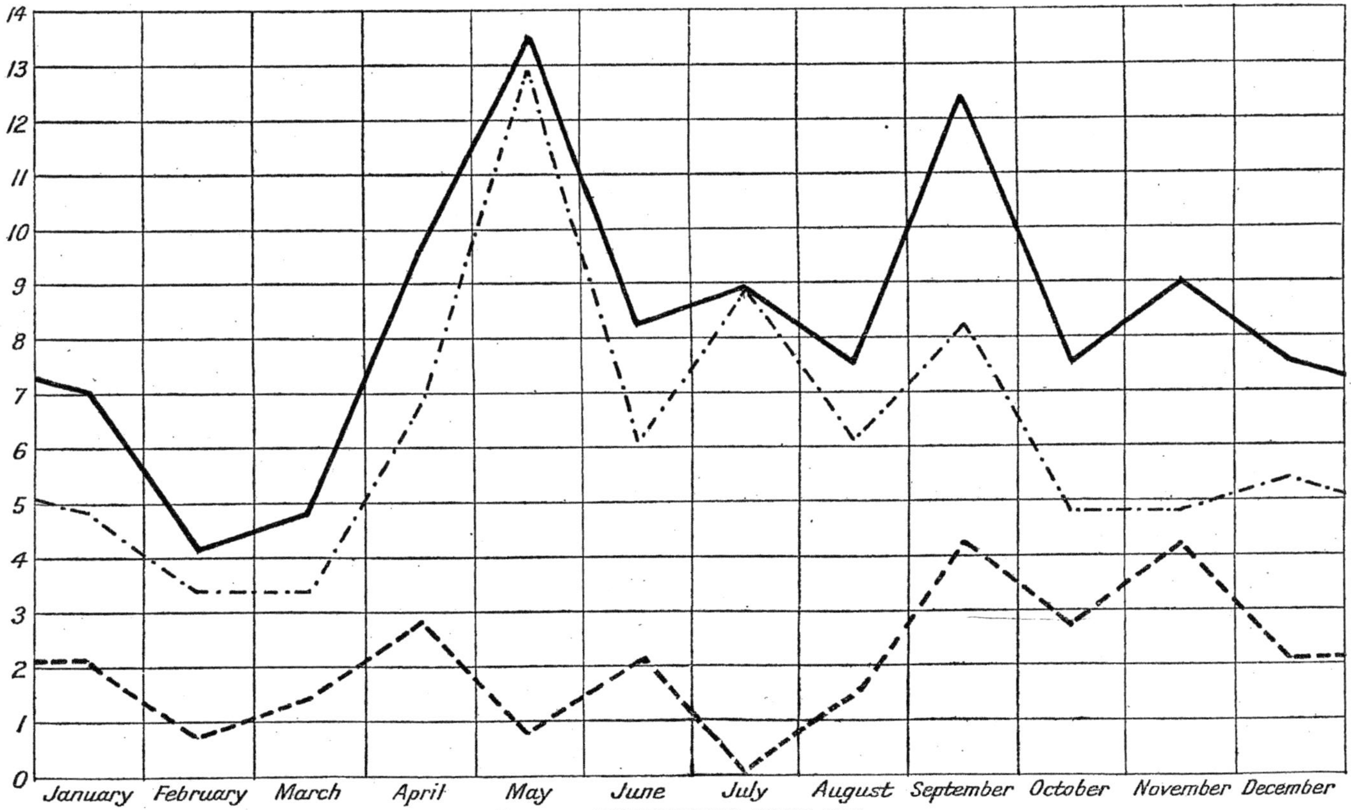

FIG. 4.—*NYCTICEBUS COUCANG*

Frequency polygons showing the monthly distribution (*a*) of all specimens in the Hubrecht Collection, Utretcht, (*b*) of pregnant specimens, and (*c*) of non-pregnant specimens. Points plotted are percentages of the total number of uteri in the collection (146).

——— All specimens ------ Pregnant specimens -·-·-·- Non-pregnant specimens deer of the family *Cervidæ* have habits as widely different as those of the Primates we are now considering, but in no scheme of classification are they more than generically separated. On the other hand, families are concerned in this distinction of breeding habit in the Primates. The variable taxonomic value of the breeding season character does not, however, seriously affect the phylogenetic significance of its differentiation among monkeys and lemurs. The subordinal separation of the Lemuroidea from the rest of the Primates is universally demanded on the strength of the distribution of morphological and embryological characters. So far as is known, breeding seasons, using the term in the sense of a demarcated season of breeding, do not occur except in this sub-order. This peculiarity of breeding habit is thus a further argument in favour of the sub-ordinal separation of this group of animals. The peculiarity, however, is obviously not a primary criterion of sub-ordinal differentiation, since it is exhibited by some, but not by all members of the group. The significance of its distribution within the sub-order could be assessed only in relation to the taxonomic criteria used in the subdivision of the group into its two series and its various families.

This peculiarity of the lemurs raises an interesting question. Have they specialized in breeding habit from a more primitive condition of continuous breeding which is still manifested by other Primates, or does the short and definite breeding season represent the primitive habit, and continuous breeding the specialized one. This question cannot be considered except in relation to certain palæontological facts concerning the geological appearance of the different primate groups. True lemuroid Primates (e.g. *Pelycodus* of the *Noth-*

arctidæ) appear first in the Lower Eocene, and according to Matthew (1928) they have no known Palæocene ancestors. In that geological horizon they are altogether differentiated from tarsioid Primates (e.g. *Tetonius*), which first appear in the Palæocene, and which are apparently connected with the Palæocene and Eocene family of tree-shrews (e.g. *Nothodectes* of the *Plesiadapidæ*). In spite of this, palæontological authorities (see Gregory, 1922) do not doubt that the primitive lemuroids of the Eocene (i.e. *Notharctidæ* and *Adapidæ*) evolved from a menotyphlous stock (e.g. *Plesiadapidæ*) either in the Cretaceous or in the Palæocene. In discussing the derivation of primate breeding habits, it is necessary, therefore, to know about the breeding habits of the Menotyphla. It is also essential to remember that phylogenetic speculation of this kind always makes far-reaching assumptions about the physiological habits of extinct animals, for in pursuing these speculations it is necessary to suppose that structural affinities between existing and fossil forms imply physiological affinities as well. Such an assumption is tacitly accepted in the zoo-geography of Palæontology. In tracing the history of a group from its derivation in the Palæocene to the present day, it is assumed that the animals belonging to the group have always reacted to climatic and food conditions as their modern representatives would do to-day,*—that is to say, that their physiological characters have remained as constant as their structural ones. For the purposes of the present argument, therefore, it may perhaps be assumed that the breeding habits of modern Primates represent those of the fossil types to which they approximate structurally.

* See Matthew, 1915; Black, 1925.

Unfortunately, little is known about the breeding habits of existing Menotyphla, and, so far as I can discover, the naturalistic literature does not contain any reference to the question. The Hubrecht (Utrecht) embryological collection of *Tupaia*, which I have discussed elsewhere in its bearing on this problem (Zuckerman, 1932*d*), does not include specimens taken in all months. It does, however, provide enough data to show that *Tupaia* breeds during at least eight months of the year. There are no data about the other four months of the year, and it is quite likely that the animal breeds continuously.

Whether or not *Tupaia* gives birth during more than eight months of the year, its breeding season is definitely longer than that of the Lemuriformes. It could thus be argued that the restriction of the breeding habit amongst the Mascarene (possibly, too, the African) prosimians is a specialization from a primitive habit of more continuous breeding, such as is still demonstrated by the primitive *Tarsius*, which is capable of giving birth throughout the year. A conclusion of this kind would be consonant with the opinions of those morphologists who hold that the lemurs are a specialized and divergent group of Primates.

(*d*) The Reproductive Cycle *

Pithecoidea.

It has long been known that non-pregnant catarrhine Primates experience a menstrual cycle, which is a rhythmically recurring œstrous cycle lasting approximately thirty days, successive cycles being separated

* The reproductive phenomena referred to in the following pages of this chapter have been discussed in detail from the physiological point of view in Zuckerman, 1930, 1932*a* & *b*.

by externally visible uterine bleeding. This type of cycle is characteristic of man, and of all the apes and Old World monkeys that have been observed. Since abundant opportunity for observation of New World monkeys and lemurs has failed to reveal the existence of such a cycle in them, there is little reason to doubt that the menstrual cycle occurs only in the catarrhine section of the Pithecoidea. It is important to note that the available data suggest that the physiological mechanisms and morphological changes of the cycle are much the same in all forms belonging to this group.

Like the Old World monkeys and apes, all the platyrrhine Primates of South America have a unilocular uterus, but beyond the fact that they must be polyœstrous, since they seem to breed at all times, practically nothing is known of their reproductive mechanisms. Vaginal bleeding of a non-menstrual kind has occasionally been observed in marmosets and capuchins, and it also appears that the *Hapalidæ* experience demarcated periods of œstrus.

Tarsioidea.

Unfortunately very little is known of the reproductive cycle of *Tarsius*. The animal has a bicornuate uterus. In the absence of pregnancy it is presumably continuously polyœstrous, successive cycles being separated by uterine hæmorrhage (see Van Herwerden, 1905). There are no data relating to the duration of these cycles, but certain observations made by Van Herwerden indicate that they resemble the menstrual cycle more than they do the typical œstrous cycle. According to this observer, ovulation in *Tarsius* occurs when there is no sign of uterine degeneration, and her evidence strongly suggests that ovulation usually precedes a period of endometrial degeneration. Such a histological picture is characteristic of the menstrual as opposed to the œstrous cycle, for contrary to the condition shown in the cycles of *Tarsius* and of catarrhine Primates, the majority of non-primate mammals whose œstrous cycles have been investigated present signs of uterine degeneration at the time of ovulation.

An important fact revealed by Van Herwerden is that the pregnant *Tarsius* may "menstruate", at least in the earliest stage of gestation.

Lemuroidea.

The literature contains no reference to the reproductive mechanisms of this group of Primates. All the animals belonging to the sub-order have a bicornuate uterus, and apparently they do not experience menstrual cycles. The Asiatic form *Nycticebus coucang* must be continuously polyœstrous in the absence of pregnancy, since it can breed at all times. It remains to be seen whether members of the genus *Lemur* are monœstrous or polyœstrous during their brief sexual season.

In spite of the wide gaps in our knowledge of the reproductive mechanisms of the Primates, it seems quite plain that these physiological processes are differently developed in the bigger sub-groups of the order. It is of interest that *Tarsius* seems to have a cycle closely similar to the menstrual cycle, and that the Old World

Primates differ, apparently widely, from those of the New World in respect to their reproductive cycles. This difference can be added to many others relating to structure, and together they go to show that the platyrrhine and catarrhine divisions are widely separated groups of the sub-order Pithecoidea. The menstrual cycle is indeed yet another character which (like the number of their teeth) man, the apes, and the Old World monkeys have in common. As such a cycle does not occur among the New World Primates, it would seem that the two divisions of the sub-order Pithecoidea diverged before the anthropoid and monkey stems of the catarrhine division had become distinct. According to the palæontological evidence this division took place not later then the Lower Oligocene, and the differentiation of the Pithecoidea into its two subgroups must therefore be supposed to have taken place at some earlier period, probably during the Eocene.

(*e*) The Changes in the Sexual Skin

Although the fundamental physiological plan of the menstrual cycle appears to be the same in different catarrhine Primates, important variations occur in its external manifestations, and these variations have taxonomic value. The changing character is the sexual skin, a specialized area of the body surface, usually the perineal region, which in many species reddens and swells, mainly during the first half of the cycle.* This peculiar change synchronizes with the first or follicular phase of the ovarian cycle. When the ovarian follicles begin to grow, the sexual skin becomes active. When the phase of follicular growth is at its height, the skin

* It may be noted that no such change is known to occur in the platyrrhine Primates.

is most swollen. When the follicle ruptures (i.e. when ovulation occurs, which is usually about the middle of the cycle), the swelling subsides. The correlation between these ovarian changes and the changes of the sexual skin is even closer than the above description suggests, for the sexual skin is activated by the follicular hormone, a secretion which is believed to be elaborated, at least partly, by the growing ovarian follicles. This close connection between the follicular hormone and the activity of the sexual skin has been demonstrated by injecting the hormone into spayed female monkeys, whose sexual skins have atrophied following the operation. Such procedure results in the swelling of the skin, and the amount and duration of the swelling are in proportion to the amount and duration of the injections.

Activity of the sexual skin varies in intensity from species to species, from individual to individual in a species, and in the same individual from time to time. From the taxonomic point of view, it is fortunate that the variations which occur within a species, and in the same individual from time to time, are not sufficiently pronounced to mask generic and specific differentiation. In a baboon the amount of cyclical swelling is rarely less than nine inches in diameter and six inches deep. On the other hand, in a common macaque the swelling, though distinct, is never more than slight.

Among the apes the sexual-skin change is known to occur in the chimpanzee. Amongst monkeys it is shown by one species (*Talapoin*) of the genus *Cercopithecus*,* by apparently all the species of the genus *Cercocebus*, by most of the species of the genus *Macaca*,

* Only one female of this species has thus far been reported on (Zuckerman, 1930).

and apparently by all the species of the genera *Cynopithecus*, *Mandrillus*, and *Papio*. A sexual-skin change of a very specialized kind (on the chest) has been reported in the genus *Theropithecus*, but no data are available to indicate whether the two species allocated to this genus also possess the commoner perineal sexual skin.

Primates which do not show the sexual-skin change include man, the orang,* and the gibbon, three forms which belong to the anthropoid group. Amongst monkeys it is absent from (*a*) members of the subgroup *Colobidæ* (*Semnopithecidæ*) of the family *Cercopithecidæ*; (*b*) all except one of the many species of *Cercopithecus* which have been under observation; (*c*) all the species of the genus *Erythrocebus*; and (*d*) apparently some species of the genus *Macaca*.

Adult female gorillas, most species of the sub-family *Colobidæ*, and some species of the genera *Cercopithecus* and *Macaca* have not been under observation. Pocock (1925*c*), however, has made an extensive survey of the perineal region in dried skins of catarrhine Primates. According to him, no trace of a sexual skin occurs in *Cercopithecus* or in the sub-family *Colobidæ*. Hartmann (1886 and 1904) states, without clearly indicating the source of his information, that the gorilla shows sexual-skin changes. Certain anatomical observations of Gerhardt (1906) seem to support this view.

It is clear that the changes in the sexual skin have taxonomic significance, and as such they demand closer examination. In the first place it may be noticed that the sexual-skin swelling is the expression of the inter-

* Aulmann and Brandes. Personal communications. Fox (1929) records the occurrence of perineal swelling during the pregnancy of an orang.

[*Photo. by Prof. W. E. Le Gros Clark*

7. SPECTRAL TARSIER. Family *Tarsiidae*

[*Photo. by F. W. Bond*

8. RED-FACED UAKARI MONKEY

action of two factors. It is the result of the action of a specific hormone on the blood vessels of an area of the body specialized in a way that is not yet understood.* The first of these two factors is a chemical one, the second is morphological, and the two do not coincide in their systematic distribution. It is a fairly safe generalization that the follicular hormone is present in the follicular fluid of all mammals, and therefore in the follicular fluid of all catarrhine Primates. But as is plain from what has been stated above, the vascular specialization is not regularly distributed through the catarrhine group. Thus two factors, the one physiological and the other structural, which together make up a morphological character of taxonomic value, are not necessarily always linked.

On the whole, the distribution of sexual-skin changes does not cut across classification by morphological characters, although if the traditional arrangement of the Old World Primates is accepted, it is quite plain that as a taxonomic character this physiological phenomenon has different value in the different sub-groups of the catarrhine division of the Primates. Thus while none of the members of the family *Hylobatidæ* shows the periodic swelling, it is variously developed among the different genera of the families *Pongidæ* and *Cercopithecidæ*. In this, however, it is not peculiar, for as Pocock has pointed out (1925*c*, p. 1570), the same is true of the morphological characters used in expressing the differentiation of the genera and species of the family *Cercopithecidæ*.

The presence of a sexual skin in the chimpanzee and

* It may be noted here that this vascular peculiarity is not limited to the perineal region. For example, the face may also be specialized, as in *Macaca mulatta*, while in *Theropithecus gelada* the skin of the chest is said to be specialized in this way.

its absence in the orang is in accordance with accepted views, based on morphological and palæontological study, of the phyletic divergence of these two types. In the family *Cercopithecidæ* it is as a rule exhibited either by all or by none of the members of the different genera that are established on the basis of the differentiation of superficial characters. Thus, to give a single example, the changes apparently do not occur in any of the genera classified in the sub-family *Colobidæ*.* In two other genera (*Cercopithecus* and *Macaca*), however, some but not all of the species show the changes. It is necessary to consider whether these genera do not require revision on the basis of the distribution of the sexual skin.

The genus *Cercopithecus* is divided into some thirty species. Only one of these, *Talapoin*, is known to experience a sexual-skin cycle. The Talapoin monkey is peculiar in other ways. It is less than half the size of any other member of its genus, and its fellow species vary in size relatively little among themselves. This difference, and a few others which it displays, inspired some systematists (see Elliot) to give it generic status.

* The removal of the *Colobidæ* into a family of their own and distinct from *Cercopithecidæ* is sometimes advocated because the *Colobidæ* (*a*) lack cheek-pouches, which are a constant character of other Old World monkeys, and (*b*) possess sacculated stomachs, which are unknown in other monkeys. Apart from these discontinuous characters, they are also differentiated from other Old World monkeys in the relative development of certain other structural features, e.g. nose, hands, feet, and in some very slight dental modifications. All these characters, however, do not imply so wide a gap between the *Colobidæ*, on the one hand, and all other Old World monkeys and baboons on the other, as those which justify the separation of the *Pongidæ* from the *Hylobatidæ*, and of these two families from all monkeys and baboons. The most they would seem to justify is the separation of the monkeys which constitute the *Colobidæ* into a sub-family of the *Cercopithecidæ*. This course was adopted by most older systematists, but is rejected in Pocock's recent classification.

This procedure has not been accepted in recent years (e.g. Pocock, 1925*b*; Schwarz, 1928*a*; Flower, 1929).

As their main reason for referring to the Talapoin by the generic name *Miopithecus*, older systematists stated that its lower third molar possessed only three cusps (the third molars of all other members of *Cercopithecus* have four). In 1925 Pocock pointed out that this is incorrect, and that the tooth actually has four cusps, and in 1928 Schwarz drew attention to the strange fact that it is the upper third molar of the Talapoin which has three cusps, and not four like most other cercopitheques. Schwarz believes that the statement that the lower tooth had three cusps was due to a *lapsus calami* on the part of I. Geoffroy Saint Hilaire, the original describer of this monkey, the error being later passed on from book to book. This actual dental peculiarity is not restricted to this species of *Cercopithecus* only, and thus Schwarz, like Pocock, does not subscribe to the view that the Talapoin should be given generic rank. Neither of these authors appears to have been aware that the Talapoin has a sexual skin. Since it is the only species of *Cercopithecus* to show such a specialized area of the body, there seems some good reason to separate it into a genus of its own, *Miopithecus*, as I have done in the chart on p. 17; certainly this character is more striking than differences which have been regarded as sufficient to establish generic rank (e.g. see Pocock, 1924). Moreover, Pocock (1925*c*) has himself drawn attention to the fact that the Talapoin, apart from being the smallest of all the species of *Cercopithecus*, retains "in the skull, at all events, certain characters found in immature specimens of the larger species."

The other genus in which the sexual-skin phenomenon does not appear to characterize all species, is *Macaca*. Here the limited state of our knowledge prevents the distribution of the sexual-skin phenomenon being made the basis for a revision of the group. Many species (e.g. *irus*) which systematists have stated do not experience sexual-skin changes, have been found, on more careful observation, to do so. Different species of this genus do show considerable variation in their expression of the phenomenon, but it is doubtful if these variations are sufficient reason to divide the group, as has been attempted before on different grounds, into sub-genera. They may be of no greater taxonomic value than the interspecific variations, discussed in the following chapter, that occur in the structure of the hæmoglobin in the genus *Papio*.

Regarded physiologically, the occurrence of sexual-skin activity in a monkey is important because it is a sign that the animal is in a phase of heightened sexual activity (œstrus). This phase, as was pointed out above, is in turn correlated with the imminence of ovulation. Amongst animals living a social life, the consequences of a specially demarcated œstrus are very great, and although in catarrhine Primates the cyclically returning sexual-skin swelling merely represents a periodic increase in the intensity of a more or less uninterrupted sexual life, the social consequences of the phenomenon are nevertheless profound (see Zuckerman, 1932*a*). There is no evidence from any quarter to tell us whether or not man's immediate sub-human primate ancestors exhibited sexual-skin activity, but the fact that there is no trace of a sexual skin in *Homo sapiens* suggests that they did not.

The distribution of the sexual skin among the various Old World Primates would appear to have little phylogenetic significance. Pocock (1925*c*), it is true, has argued that the similarities of its manifestations in *Cercocebus* and *Papio* suggest the close kinship of these two groups, a view also supported by the fusion of the anal callosities in the males of these two genera. The seemingly chaotic distribution of the character among the Catarrhini, and in particular the variety of its expressions in the genus *Macaca*, seems, however, to indicate that it has developed independently among different Primates. If such a view is denied, it would be necessary to adopt a diphyletic, and perhaps a polyphyletic, scheme to explain the relationships not only of the genera of monkeys but also those of the apes.

CHAPTER V

THE DIFFERENTIATION OF BLOOD REACTIONS

THE characters of the blood of the Primates may be discussed from the points of view of taxonomy and phylogeny under the headings of (*a*) the morphology of the blood cells, (*b*) the crystallography of the hæmoglobins, (*c*) the serum precipitin reaction, (*d*) the blood groups, and (*e*) the specificity of the red cells. In view of the unanimous admission of the relevance of hæmatology to the problems of animal classification, it is unnecessary to introduce this chapter with any general discussion. As I hope to show, the facts to a great extent speak for themselves.

(*a*) The Blood Cells

Ponder, Yeager and Charipper (1928, 1929) have recently published a report of an extensive investigation on the blood cells of the Primates. Their main results are summarized in table II, opposite. It is unnecessary here to refer in detail to their observations, but in passing it may be noted that, with the exception of the green monkey, the marmoset, the capuchin and the ring-tailed lemur, the red blood corpuscles of all the species investigated appear to be more resistant to hypotonic saline than are human red blood corpuscles —those of the common macaque showing the greatest resistance. The behaviour of the blood corpuscles to other hæmolysins (e.g. sodium taurocholate and saponin)

TABLE II

Hæmatological Characters of the Primates, from Ponder, Yeager and Charipper

Species.	Red Cells per c. mm.	White Cells per c. mm.	Hæmo-globin. Per Cent.	Mean Diam. R.B.C. in Plasma.	Differential Count.					
					P.M.N.[12]	P.M.E.[13]	P.M.B.[14]	L.[15]	L.M.[16]	T.[17]
Man	5,400,000 [1]	7,000	105	7·9 (Ponder)	67	2	1	25	5	—
Gorilla	6,250,000	6,800	83	7·7	63	5	3	23	4	2
Chimpanzee	7,300,000 [2]	10,400	89	7·8	58	5	20	16	1	—
Orang	6,880,000	9,400	80	7·8	55	4	15	24	2	—
Yellow Baboon	6,970,000	10,400	87	7·7	65	2	1	29	2	1
Green Cercopitheque	6,400,000	12,600	87	7·8	58	7	1	31	3	—
Moor Macaque	5,000,000	7,600	88	7·9	69	2	4	23	1	1
Rhesus Macaque	5,000,000	10,400	77 [3]	8	73	3	1	18 [4]	2	—
Common Macaque	6,432,000 [5]	7,200	90	8	37	19	24	18	1	1
Brown Capuchin	5,100,000 [6]	10,400	90 [7]	7·8	68	5	3	21	2	1
Black Spider Monkey	5,760,000	10,000	76	9·1	69	12	1	18	—	—
Geoffroy's Spider Monkey	3,840,000 [8]	7,000	80	8·8	73	8	3	15	1	—
Squirrel Monkey	7,416,000	11,000	84	6·4	65	6	2	26	1	—
Douroucouli	4,664,000	8,200	71 [9]	7·1	79	8	1	12	—	—
Marmoset	6,624,000	7,800	67 [10]	7·7	72	2	4	19	3	—
Ring-tailed Lemur	7,936,000	16,400	87	6·8	66	7	2	23	1	1 [11]
Brown Lemur	10,304,000	15,400	75	6·7	69	1	3	27	—	—

[1] The figures for man are those usually taken as an average in clinical studies. The figures for red blood corpuscles and hæmoglobin concentration are taken from Price-Jones (1931), and are figures for males; the diameter of corpuscles from Ponder (1924).

[2] Given as 7,300,000 in *Zoologica* (1929), and as 6,300,000 in the *Quart. J. Exp. Phys.* (1928).

[3] ,, 77 ,, ,, ,, 89 ,, ,, ,,

[4] ,, 18 ,, ,, ,, 21 ,, ,, ,,

[5] ,, 6,432,000 ,, ,, ,, 6,434,000 ,, ,, ,,

[6] ,, 5,100,000 ,, ,, ,, 4,700,000 ,, ,, ,,

[7] ,, 90 ,, ,, ,, 78 ,, ,, ,,

[8] ,, 3,840,000 ,, ,, ,, 4,840,000 ,, ,, ,,

[9] ,, 71 ,, ,, ,, 75 ,, ,, ,,

[10] ,, 67 ,, ,, ,, 77 ,, ,, ,,

[11] ,, — ,, ,, ,, 1 ,, ,, ,,

[12] Polymorphonuclear neutrophil leucocytes.

[13] Polymorphonuclear eosinophil leucocytes.

[14] Polymorphonuclear basiphil leucocytes.

[15] Lymphocytes.

[16] Large mononuclear leucocytes.

[17] Transitional leucocytes.

is not quite the same, little relation being found between resistance shown to one lysin and that shown to another.*

(*b*) The Crystallography of the Hæmoglobins

Reichert and Brown (1909) have shown, by crystallographic methods, that the globin or proteid part of hæmoglobin is different in different vertebrates (the hæmatin fraction is the same in all animals [Anson, Barcroft, and co-workers, 1924]). Their investigations, so far as the Primates are concerned, were limited to *Lemur catta*, the ring-tailed lemur; three species of the genus *Papio* (*cynocephalus*=*babuin*, the yellow baboon —in which, following Elliot, I include here the species *langheldi*, although Reichert and Brown regarded it as a distinct type; *porcarius*, the Chacma baboon; *anubis*, the olive baboon); both the drill (*leucophæus*) and mandrill (*sphinx*) of the genus *Mandrillus*; and man. With the exception of the yellow baboon, apparently only one specimen of blood was examined in each of the sub-human species.

These investigators have shown that the oxyhæmoglobin crystals of the genera *Papio* and *Mandrillus* belong to the same crystallographic system (i.e. are isomorphous), and they were able to demonstrate three types of crystal, except where the material was insufficient or bad. Two of these types also occurred in human blood, and were strikingly similar to the baboon crystals, both being orthorhombic, to which system the crystals of the blood of the lemur also belonged. Monoclinic oxyhæmoglobin crystals were obtained only from the several species of baboons, which differed amongst

* See Solowiev, 1930, for a short paper on the specific weight of the blood of monkeys.

themselves with regard to the angles and ratios of the crystals.

Although their taxonomic importance has often been emphasized (e.g. see Robson, 1928), it is unnecessary here to refer in greater detail to Reichert and Brown's observations. In the case of the Primates they are too fragmentary to do more than point the way to what may prove a fertile field of phylogenetic research.

(*c*) The Serum Precipitin Reaction

The researches on the blood serum precipitin reactions of Primates, initiated in England early in this century by Grünbaum (1902) and by Nuttall (1904), are too well known to demand more than brief mention here. Their experiments have been repeatedly confirmed, and have also been extended by many other workers. The technique of this reaction is quite simple. Rabbits are immunized against human, ape, and monkey sera, and the anti-human, anti-ape, or anti-monkey sera so obtained are then tested with the sera of other Primates, to see to what extent positive precipitin reactions are obtained. The amount of precipitation is also recorded.

A paragraph from Nuttall's book (1904, p. 214) excellently describes the results yielded by these experiments.

"These tests were conducted by means of antisera for man (825 tests), chimpanzee (47 tests), ourang (81 tests), *Cercopithecus* (733 tests). Maximum reactions were only obtained with bloods of Primates. The degrees of reaction obtained indicate a close relationship between the *Hominidæ* and *Simiidæ*, a more distant relationship with the *Cercopithecidæ*, the bloods of *Cebidæ* and *Hapalidæ* giving still smaller reactions than the last, when we consider the results obtained with the first three antisera. The tests with antiserum for *Cercopithecus* gave the largest reactions with bloods of *Cercopithecidæ*, next with those of *Hominidæ* and *Simiidæ*, but slight reactions with those of *Cebidæ* and *Hapalidæ*. All four antisera failed to produce reactions with the two bloods of *Lemuridæ* tested, except when sufficiently powerful to also produce reactions with other mammalian bloods. From this we may conclude that the Lemurs properly belong to an Order separate from the other Primates."*

The last sentence in this paragraph implies an unduly strong belief in the taxonomic value of the precipitin reaction. That it is unsafe to lay so much taxonomic emphasis on this test is suggested by the fact that *Tarsius*, which undoubtedly *is* a Primate, gives a negative reaction, as does also *Nycticebus* (Le Gros Clark, 1924*a*).

Nuttall and Strangeways also attempted a quantitative expression of the relationships of the Primates, by estimating the volume of precipitate, a method not altogether free from objection (see Landsteiner and Miller, 1925*a*, and Boyden, 1926). Using anti-human serum the following relative figures were obtained:

Man	100
Chimpanzee . . .	130 (precipitate less compact)
Gorilla	64
Orang	42
Mandrill	42
Guinea baboon . .	29
Spider monkey . .	29

With anti-orang serum the figures were:

Orang	100
Man	75
Rhesus monkey . .	62

* Nuttall tested the following catarrhine bloods: the gorilla, chimpanzee, orang; various species of the genera *Papio, Mandrillus, Theropithecus, Macaca, Cercopithecus, Presbytis*. And the following platyrrhine bloods: species from the genera *Alouatta, Cacajao, Aotes, Saimiri, Ateles, Lagothrix, Cebus, Hapale*. The only lemur bloods he tested belonged to the family *Lemuridæ*. The most recent general corroboration of his conclusions is that of Wolfe, 1933.

Gibbon serum was not tested in these investigations, but has since been examined by Le Gros Clark (1924*a*), who notes that it gives a positive reaction with anti-human sera, and by Mollison,* who remarks that the amount of reaction he observed was almost equal to that of the chimpanzee. This observation will be referred to again below. From the point of view of the affinities of the various members of the order Primates, the figures given above speak for themselves.

(*d*) The Blood Groups

As is now well known, the blood serum and red blood corpuscles of the apes, including the gibbons, respectively contain group iso-agglutinins and group iso-agglutinogens which are specifically the same as those found in human blood. These cannot be demonstrated directly with human serum, since the latter also contains hetero-agglutinins for the corpuscles of the apes, the effect of which has to be obviated before it is possible to demonstrate the presence in these animals of the blood groups which are identical with those of man.†

Many investigations of this peculiar man-ape relationship have been made, all of them being limited on account of the regrettable lack of material. Recently both Weinert (1932) and Lattes (1932) have collected

* *Vide* Weinert, 1932.

† For full discussion see Landsteiner and Miller (1925*b*) and Snyder (1929). Group specific iso-agglutinin solutions for this work are prepared by absorbing the iso-agglutinins of human serum with human corpuscles, from which they are then freed by heat. In this way solutions are prepared free from hetero-agglutinins. Another method of preparing group specific absorbed immune sera is to use rabbits, first immunizing the animals with human cells of different groups, and then absorbing the hetero-agglutinins of the rabbit serum with group O human cells.

together the published records of these investigations. The table given below is to some extent a synthesis of the figures they provide, since each author seems to have overlooked data considered by the other. To their figures have also been added the records of two orangs (published by Penrose, 1932) which neither of them mentions.

TABLE III

Blood Grouping of *Hylobatidæ* and *Pongidæ*

	O	A	B	AB
Chimpanzee * . .	6	58	—	—
Gorilla	—	4	—	—
Orang †	—	4	6	3
Gibbon	—	2	6	2

* Lattes gives A 58, O 6. Weinert A 56, O 6.

† Lattes mentions only ten orangs, Weinert eleven. To the latter's eleven have been added Penrose's two.

The investigation of the blood groups has naturally not been limited to the apes, and very interesting results were obtained when lower Primates were tested. The present account is taken from a paper of Landsteiner and Miller (1925*c*). These authors examined the following catarrhine monkeys: eight animals belonging to the genus *Papio*; ten belonging to the genus *Cercopithecus*; one belonging to the genus *Erythrocebus*; four belonging to the genus *Cercocebus*; twenty-three belonging to the genus *Macaca*. Twenty-two New World monkeys were also examined, nineteen belonging to six different genera of the family *Cebidæ*, and three to one genus of the family *Hapalidæ*. In addition to these, eight lemurs were examined, all belonging to the family *Lemuridæ*.

All these animals were tested with group specific sera prepared in the way briefly summarized in the footnote to p. 53. Not one of the forty-six Old World monkeys thus examined gave any indication of a true reaction, and most of them gave entirely negative responses.* On the other hand, all the New World monkeys, and some of the lemurs, gave distinctly positive reactions with an agglutinin solution Group II. It was possible to state in the case of the apes that the iso-agglutinogens of their erythrocytes are specifically the same as those of human red blood cells. This means that after being saturated with the agglutinogen from the cells of an ape, a serum is incapable of agglutinating human cells containing the same agglutinogen.† This applies to sera prepared by both methods mentioned in the footnote to p. 53. The New World monkeys and the lemurs, however, which, by reacting to Group II human purified agglutinin solution, attested to the presence

* Thomsen and Kemp (1930) tested two baboons and fourteen macaques (seven belonging to the species *M. mulatta*) with the same sera used in typing human blood (titre for human B cells 1/128), and they suggest that the red cells of nine of these animals contain B receptors, their observations thus conflicting with those of Landsteiner and Miller, who failed to find an agglutinogen B in the blood of Old World monkeys. Of the nine positive bloods, two agglutinated in titres no higher than 1/4, five in 1/8, and the remaining four in titres no higher than 1/16. Earlier in their paper Thomsen and Kemp argue that titres for rabbit cells in human anti-A serum of the range from 1/8 to 1/32 (titre for human cells 1/128) are representative of hetero-agglutination, and not group-specific agglutination. They themselves thus throw doubt on their conclusion that the cells of Old World monkeys contain the receptor B, since the figures they obtained with the bloods of the monkeys may also have been representative of a hetero-agglutination response. In any case, the capacity of the monkey cells to absorb the anti-B agglutinin cannot be very high, since absorption experiments with four bloods in various concentrations showed that the anti-B sera could still agglutinate human cells in titres of from 1/16 to 1/64. Further investigation is necessary to decide between the conclusions of these observers and those of Landsteiner and Miller.

† See Landsteiner, 1928, and Troisier, 1928.

of an agglutinogen related to B,* nevertheless failed to react to absorbed group specific immune sera prepared by the use of rabbits. This suggests that their agglutinogen B is not specifically the same as that of man and the apes. Not surprisingly, therefore, the blood of these Primates fails to absorb all the human iso-agglutinins contained in the serum to which they actually do react.

These facts appear to be of very great phyletic importance. The uniformity of the results of Landsteiner and Miller's experiments indicates, as they write, "a definite rule", and the blood-group character exposes sharp limits to the different primate families, there being no gradual transition from one group to another. Like the distribution of the menstrual cycle, that of this hæmatological character indicates a wide phyletic gap between the monkeys of the Old World and those of the New. It also suggests a wide gap between the members of the family *Cercopithecidæ* on the one hand and the apes and man on the other. But before discussing the significance of the presence of specifically the same agglutinogens and agglutinins in all members of the anthropoid group of Primates, certain other facts relating to the characters of the blood require mention.

(*e*) The Specificity of the Red Blood Cells of the Primates

In the light of the evidence of the blood groups and precipitin reactions, it would seem that, except for the presence in each of reciprocal hetero-agglutinins, there is little hæmatological differentiation between

* Such an agglutinogen is shown by many animals.

man and the apes. By the use of anti-erythrocyte sera, however, differentiation has been revealed in the red cells, and this method, as Landsteiner and Miller (1925*a*) write, "will regularly differentiate between species so closely related that their sera are indistinguishable by means of the precipitin reaction." It was in a somewhat similar way that Marshall (1901–5) showed that while anti-human erythrocyte sera caused as much hæmolysis of macaque as of human blood, similar anti-macaque sera had little hæmolytic effect on human blood.

This method of investigation has been well exploited by Landsteiner and Miller, who found that with anti-human erythrocyte sera prepared by the use of rabbits, there is little, if any, difference in intensity of agglutination between man, the chimpanzee, and the orang (one experiment); the amount of agglutination was less with gibbon blood (one experiment). More striking results were obtained by absorbing anti-human sera with chimpanzee red blood cells, and anti-chimpanzee sera with human red blood cells. After such procedure it was found that these sera still contained agglutinins for human blood and for chimpanzee blood respectively. This method thus provides a means of differentiating between closely allied bloods. By its use it was possible to show (*a*) that orang blood is more closely related to chimpanzee than to human blood; (*b*) that human and chimpanzee bloods have a greater affinity with each other than chimpanzee blood has with the blood of baboons and macaques; (*c*) that the latter two can be differentiated from each other; and (*d*) that both are conspicuously different from the blood of a New World monkey (*Cebus hypoleucus*). Anti-macaque and anti-baboon sera absorbed with the red cells of this monkey still had full agglutination power for macaque and baboon erythrocytes.*

The Phyletic Bearings of the Blood Reactions

The first point that may be noticed about the reactions described in the preceding pages is that they do not always point to the same phyletic conclusion. Thus, to give an example, gibbon serum gives a greater precipitate with anti-human serum than does the serum of the orang, but it is the blood of the orang which appears to be more "human" in its reaction to anti-human erythrocyte sera.

It is difficult to differentiate between the values of the various reactions, but there can be little doubt that greater emphasis may be laid, for practical purposes and from the point of view of phylogeny, on those characters which appear to be discontinuous (i.e. those that are limited to natural sub-groups), than on those that vary in a continuous and orderly way throughout all groups of Primates. Thus in their bearing on the question of man's descent, the existence in apes alone of blood groups specifically the same as those of man is more significant than the precipitin responses of these animals, since these vary greatly, and even approximate to those of catarrhine monkeys. The fact of the anthropoid blood groups is of equal importance phyletically with the fact that man and the apes are among the very few mammals known to be incapable of carrying the oxidation of purine bases as far as allantoin. This metabolic process is possible for "the monkey", whereas

* Uhlenhuth (1901, 1902, 1926) has described a method of differentiating primate bloods by the precipitin reaction, and Landsteiner and Levine (1932) have recently described experiments on the immunization of chimpanzees with human blood. These further methods of differentiating primate blood are not relevant to the argument of this chapter.

in man, the chimpanzee, and the orang the process of purine metabolism ends with uric acid.*

While the serum precipitin reaction, which concerns the specificity of serum proteins, suggests the existence of a wide gap between the Lemuroidea and Tarsioidea on the one hand, and the rest of the Primates on the other, it seems to indicate that the latter are all in some way related, and that their human affinities become closer as one passes from the New World monkeys to those of the Old World, and from the latter to the apes. The tests with anti-human or anti-sub-human primate erythrocyte sera, which concern the specificity of the red cells, help to define these affinities, and indicate that the hæmatological gap between Old World monkeys and man, though great, is not much greater than that between the monkeys and the anthropoid apes. They might also be taken to indicate that this gap is possibly less than that existing between the blood of Old and of New World Primates. The existence of the four blood groups amongst the apes is an indisputable sign of their phylogenetic anthropoid status, and it is a criterion which cuts them off sharply from the catarrhine monkeys. It

* A general account of this interesting observation is given by both Pryde (1931) and Rose (1923). The actual investigations were carried out by Wells (1909), Wiechowski (1912), Hunter and Givens (1912, 1914), Wells and Caldwell (1914), Hunter and Ward (1920). In all, one orang, four chimpanzees, twenty-three Old World monkeys of three different genera, and one New World monkey appear to have been examined. The apes were exceptional in so far as their tissues lacked uricolytic enzymes and their urine was free of endogenous allantoin. Certain peculiar facts relating to the orang and to the New World monkey suggest the possibility of more detailed differentiation of purine metabolism in the Primates than is expressed by the statement "apes alone fail to oxidize uric acid". The whole question, like almost all aspects of the physiology of the sub-human Primates, requires further examination.

may again be noted here that the occurrence of an agglutinogen similar to B, but not specifically the same, in New World monkeys and lemurs, is no indication of their close affinity to the apes and man, since a similar factor is found amongst non-primate mammals.

Although these considerations point to the close similarity of human and ape bloods, it must of course be remembered that the two are nevertheless specifically different, as is evidenced, for example, by the existence of reciprocal hetero-agglutinins in the different species concerned.

The blood groups of the apes present an interesting problem. The human groups, on Bernstein's (1925) hypothesis, are supposed to have arisen at some time as mutations of the recessive R genes,* and writers have speculated whether this occurred during man's simian phase or during a phase when he would have been recognized as human. Many investigators (see Snyder, 1929) hold the view, which they base on the distribution of the groups amongst the peoples of the world to-day, that the A and B mutations in the R gene occurred independently in different parts of the globe and at different times during man's human phase. Others (see Woollard, 1932) hold that man inherited his blood groups from ape precursors. Table III, however, definitely seems to support the view that the same blood-group mutations occurred several times and independently in what are clearly allied

* On Bernstein's triple allelomorph hypothesis, the genetic formula of the groups may be written:

Group.	*Genes.*
O	RR
A	AA or AR
B	BB or BR
AB	AB

stocks. Thus it is significant that although only ten gibbons and thirteen orangs have thus far been tested, it is already known that the A and B mutations have occurred in both species. On the other hand, not one of sixty-four chimpanzees have given any indication of the presence of B, which suggests strongly that the species is characterized by A alone. It is of course conceivable that an original chimpanzee stock was characterized by both groups, of which B, owing to some chromosomal alteration, was later lost. Whether or not this may be the true explanation, it does not alter the fact that the blood group characters have evolved independently and differently in the African and in the Eastern apes. Since this is the case, there can be little reason for denying the possible occurrence of similar independent mutation in the homologous human genes. If, nevertheless, this is denied, and if it is insisted that man inherited his blood groups from ape or ape-like precursors, it would have to be assumed that many human beings have greater affinity with the existing Asiatic anthropoids than with those of Africa, since the latter give evidence of the presence of only factor A. This attractive view, while it can already claim supporters in virtue of some morphological considerations, is highly debatable, and would be denied by the majority of students of the subject. In the existing state of our knowledge it would be safer to support the view of independent mutations.

CHAPTER VI

THE DIFFERENTIATION OF RECEPTOR ORGANS AND THEIR FUNCTIONS

THIS chapter concerns the specific differentiation of the perceptual processes of the Primates. Unfortunately this is a field which has been studied very little as yet, and available knowledge is limited almost entirely to the functions of vision and smell.*

The Visual Processes of Primates

Macular Vision.

Experimental data relating to this subject are scanty, and refer almost entirely to the apes and catarrhine monkeys. Much, however, is known about the structural characters of the ocular apparatus of almost all Primates, and this information allows fairly safe inferences about function.

Lemuroidea.

Like those of most sub-primate monodelphian mammals, the eyes of the members of this sub-order, even of the short-snouted Galagos, are set more to the sides

* In his new book, *Behavior Mechanisms in Monkeys*, Klüver (1933) reports on the reactions of monkeys in experiments demanding the differentiation of a variety of auditory stimuli. I am greatly indebted to Dr. Klüver for the privilege of reading the proofs of this book. The opportunity to do so came when my own book was also in proof, and unfortunately I was therefore unable to add in the following pages more than a few scanty references to his important study.

than to the front of the head, the visual fields overlapping very little. Fovea and macula are absent from the retina, which conforms to what is known as the nocturnal type. (It is interesting that the retina of *Tupaia* of the Menotyphla is of the diurnal type. It too lacks a fovea and macula.)

From these facts it is safe to assume that the vision of the lemur is not stereoscopic.

Tarsioidea.

In *Tarsius*, according to Woollard (1925), "though the orbits are inclined in their major axes to each other at an angle of about ninety degrees, this is not repeated in the eyes themselves for they face almost directly forwards in the same plane." The retina of this animal, like that of the lemur, is of the nocturnal type, but differs in possessing a differentiated area lateral to the entrance of the optic nerve, the possession of which, Woollard believes (1927), enhances visual acuity. Nevertheless, the eyes of *Tarsius* are incapable of conjugate movement—the animal does not possess "the complex nervous mechanisms for muscular cooperation necessary to bring the images of an object into the corresponding positions on the two retinae" (Elliot Smith, 1926).

Like the lemur, therefore, *Tarsius* almost certainly lacks stereoscopic vision.

Pithecoidea.

The forward position of the eyes of the monkey, of the ape, and of man, allows the two visual fields to overlap greatly, and in their movements the eyes are automatically conjugated. With one known generic exception (*Aotes*) (see Woollard, 1927) apparently all of the higher Primates have a retina of the diurnal type, together with fovea and macula. The members of the Pithecoidea are perhaps the only mammals with stereoscopic vision, but although apparently capable of form, size, and distance discrimination, it is doubtful whether these abilities are always employed as they are in man. (For brief reviews of this subject see Yerkes and Yerkes, 1929, and Zuckerman, 1932*a*.)

The exceptional genus of the sub-order Pithecoidea, *Aotes*, is a group of primitive nocturnal South American animals belonging to the family *Cebidæ*. While the axes of their eyes are the same as in other monkeys, these animals have a retina of the nocturnal type, without cones, fovea or macula. Because of its cranial and dental characters, Gregory (1922) believes that *Aotes* is among the most primitive and "tarsioid" of existing Platyrrhines, and since it is generally accepted that this group of animals evolved from small Primates related to the Tarsioidea and Lemuroidea, it is possible to regard the peculiar retina of *Aotes* as the retention of a prosimian character. Woollard's alternative explanation (1927) of the existence of this single retinal exception in the sub-order Pithecoidea—that the animal's nocturnal habits provide the explanation of its retinal defects—raises a very controversial and wide issue of biology, and is incompatible with the more usually accepted views on the methods of evolutionary change.*

Colour Vision.

The facts revealed by colour discrimination experiments also indicate a wide gap between the Lemuroidea

* Many of Woollard's observations have been corroborated by Kolmer, 1930.

[*Photo. by F. W. Bond*

9. SQUIRREL MONKEY. Family *Cebidae*. Genus *Saimiri*.

[*Photo. by F. W. Bond*

10. LION MARMOSET. Family *Hapalidae*

and the higher Primates. The chimpanzee (Kohts, 1923), macaques (Watson, 1909; De Haan, 1925 *a* and *c*; Trendelenburg and Schmidt, 1930), the Chacma baboon (Zuckerman and Wallace, 1932), and the capuchin (Watson, 1909) have been investigated, and it has been clearly shown that these animals discriminate colours probably as well as we do.* On the other hand, recent experiments carried out by De Haan and Frima (1930) suggest strongly that the lemur is colour-blind. It appears to discriminate between coloured discs mainly on the basis of differences in brightness. One of the two animals (out of four) which yielded results in these experiments did, however, react in such a way as to suggest that it was capable of colour discrimination of a very feeble kind.

THE OLFACTORY PROCESSES OF PRIMATES

Experimentally controlled facts regarding this receptor system are remarkably difficult to obtain, even in the case of man, and it is not surprising that the physiology of the olfactory processes of the sub-human Primates has not been investigated. Certain inferences, however, may be drawn from the facts of comparative anatomy.

One of the striking characteristics of the primate brain, as Elliot Smith has pointed out (1927), is the reduction in the part of the cerebral hemisphere concerned with smell, and the corresponding increase in the cortical representation of visual functions. The reduction in the olfactory part of the brain is more pronounced in *Tarsius* than in any of the Lemuroidea, and is still more conspicuous in the monkeys and

* Klüver (1933) presents data which strongly suggest that the squirrel monkey (*Saimiri sciurea*) is endowed with colour vision.

apes. This suggests that the sense of smell is of greater importance to lemurs than to *Tarsius*, to whom, again, it is probably more important than to members of the Pithecoidea.

It is difficult to relate these inferences to the distribution of cutaneous scent glands among the Primates.

Amongst Lemuroidea, arm glands are found in the ring-tailed lemur, *Lemur catta*, and the grey gentle lemur, *Hapalemur griseus*; anal glands are also commonly found, especially in the genus *Lemur* (Pocock, 1918). Amongst the *Hapalidæ* complex scent glands are found in the pubic region, on the genitalia, or in the perinæum (*Œdipomidas*, *Hapale*) (Wislocki, 1930), while amongst the *Cebidæ* sternal glands have been found on the chest in *Ateles* (Wislocki and Schultz, 1925). No glands have been reported in the family *Cercopithecidæ*, but sternal glands have been found on the chest of both the orang (family *Pongidæ*) (Wislocki and Schultz, 1925; Schultz, 1921) and the gibbon (family *Hylobatidæ*) (Pocock, 1925*c*).

There can be little doubt that these glands are scent organs, but without experimental knowledge it is clearly impossible to diagnose their precise function, or to determine whether they demand for the perception of their emanations a specially developed olfactory receptor system. It is interesting to note, however, that their development is not correlated with that of the olfactory cerebral centres, which progressively diminish in morphological importance in a series beginning with the Lemuroidea, and passing through *Tarsius* to the pithecoid Primates.

CHAPTER VII

THE DIFFERENTIATION OF BEHAVIOUR PATTERNS

ANIMAL behaviour shows almost as much differentiation as does either animal morphology or physiology. Among the Primates differentiation of this kind is easily observed, and its taxonomic implications are in close harmony with classification based on external characters. Thus the behaviour of one New World monkey of the family *Cebidæ* is far more like the behaviour of another member of the same family than like that of a baboon. Similarly the behaviour of an animal of one species of *Cercopithecus* resembles that of a member of another species of the same genus more than it does that of a macaque. This clearly suggests that forms of behaviour, provided that they can be defined, would prove to be valuable taxonomic characters. They are also characters which readily lend themselves to experimental study.

In the space of a short chapter it is clearly impossible to do more than indicate certain broad differentiations in the activities of Primates, and for this purpose I have chosen a few well-defined behaviour patterns, of social significance, which will be considered in the way other characters have been treated in previous chapters.

Facial Movements

The nose and upper lip of animals belonging to the sub-order Lemuroidea are constructed differently from those of higher Primates, and in their structure they represent the primitive mammalian naked glandular rhinarium. Pocock (1918) has given full value to this fact by separating the Lemuroidea from the Tarsioidea and Pithecoidea into what he terms a "*Grade Strepsirhini*"—the remaining Primates forming a "*Grade Haplorhini.*" * It is particularly noteworthy that *Tarsius* falls into the latter group, its nose and relatively free upper lip being constructed on very nearly the same plan as those of all the Pithecoidea.

Drinking

These sub-ordinal structural differences are reflected in the facial movements which characterize the different groups. Thus lemurs drink by lapping. (So, incidentally, does *Tarsius*, in spite of its "Haplorhine" lip.) On the other hand, members of the Pithecoidea drink in the way man does, by making use of freely movable lips and cheeks.

Facial Expression

Lemuroidea. The distinction between the three sub-orders of Primates is well defined in the matter of facial expression. Here lemurs are notoriously blank. Neither when they withdraw in fear, nor approach with interest (for example, when offered food), do their cheeks and lips move. What changes of expression they manage to convey to the human observer appear

* The terms Strepsirhini, Platyrrhini, and Catarrhini were apparently first used by E. Geoffroy St. Hilaire.

to be effected entirely by flinching (which may involve the whole body) movements of the eyes and upper eyelids, and sometimes very slight movements of the forehead and ears. Mouth movements as a rule are not made except when the animals are actually engaged in one of the processes of eating, lapping, licking, grooming or fighting.

Tarsioidea. The few available notes about *Tarsius* show that in this respect it differs greatly from the lemurs. "On approaching it in its cage," writes Cuming (1838), "it fixes its large full eyes upon the party for a length of time, never moving a muscle; on drawing nearer, or putting anything near it, it draws up the muscles of the face similar to a monkey, and shows its beautiful sharp regular set teeth." This observation is corroborated by Le Gros Clark (1924*a*).

Pithecoidea. Monkeys and apes are capable of an amazing variety of facial expressions, in which the facial muscles, the jaws and the tongue take part. A human observer can soon learn to interpret their various facial movements. Some expressions (for example that which is usually termed chattering) are common to all members of the Pithecoidea. Others, again, seem to be peculiar to one kind of animal. In spite of such variations, an observer with even moderate experience of these animals will usually have no difficulty in understanding the significance of the facial expressions of almost any monkey or ape. Monkeys themselves seem to have no specific or generic barriers to this kind of understanding.

In its fullest manifestation chattering comprises rapid, rhythmical, lip, jaw and tongue movements. The jaws and lips quickly open and shut, while the tongue is rhythmically extruded, the latter movement being partly responsible for the chatter. When the tongue movements are suppressed these activities are often referred to as smacking of the lips. The manner in which the performance is executed varies from species to species, and from one social situation to another. In its complete form it is well exhibited by young macaques, where it often seems to signify a friendly greeting tinged with fear. Mouth movements of this kind are indeed an integral part of all friendly and sexual activities, which they both precede and accompany. By making them himself, a human observer can readily evoke them from an amiably disposed monkey. In modified form smacking of the lips is also exhibited by the chimpanzee, who, when grooming a fellow ape or a friendly human being, will slowly bring its lips together with a slight smack, at the same time making a louder noise as its molars come into opposition and grind together. Chattering is also manifested by marmosets, in whom it consists mostly of up-and-down jaw movements, and by all the commoner members of the *Cebidæ*. Amongst the latter the lips are very mobile, but their mouth movements can nevertheless be readily resolved into three main forms. The first is an exaggerated pout, the lips pushing forwards to make the aperture of the mouth circular. The second is the exposure of the teeth by the drawing back of the lips. The third is the rhythmical up-and-down movement of the jaws—"chattering". The sounds of these animals are very varied, and capuchins, in particular, appear to vocalize with ease even when their lips are shut.

Exhibiting the teeth in a wide yawn in moments of dominance preceding fights—a movement which may justifiably be interpreted as a threatening one—is another expression common to apparently all the catarrhine Primates. It appears to have slight variations, as, for example, the complete eversion of the upper lip of the Gelada baboon.

The facial movements, accompanying different emotional conditions, which are peculiar to the chimpanzee, are referred to at great length by Yerkes and Yerkes (1929), and therefore need not be considered here. A curious specific expression is that of the pig-tailed macaque. This animal erects the hairs of its scalp, which it draws back; at the same time it wrinkles and draws up its face and slightly everts its upper lip. These facial movements may be made more significant in expression by the animal depressing its shoulders and raising its head. The whole complex of movements usually implies a kind of dominant curiosity, and it often precedes sexual or aggressive activity. As a last example of specific facial expression amongst the Catarrhines may be mentioned the extrusion of the lips into a small funnel, exhibited by members of the genus *Cercopithecus*, e.g. *C. leucampyx*, the Pluto monkey. This gesture seems to be a friendly and slightly suppliant one.

Grooming

Lemuroidea. The activities comprising grooming behaviour differ considerably amongst the Primates. Teeth, tongue, hands and feet are all made use of by members of the Lemuroidea. Thus if a group of animals belonging to the genus *Lemur* is watched for any length of time, the following grooming activities will be observed:

(*a*) Licking, and raking through the fur with the procumbent lower incisors. The latter are said to be cleaned by the sub-lingua, to which Pocock (1918) refers as the tooth-brush. This manner of grooming is employed for apparently all accessible parts of the body. The hair of slightly inaccessible areas, e.g. towards the rump, may be made more accessible by pulling with both hands.

(*b*) Other parts of the body, particularly the sides, are cleaned by scratching with the feet, the left foot being used for the left side, the right for the right. As Pocock has pointed out, only the second digit of the hind foot (which is short and has a short terminal pad and a long, semi-erect claw) functions as a scratcher.

(*c*) Mutual grooming is often seen, one animal using its incisor comb and tongue on the other. The hands of the grooming animal are used only to hold on to the fur of the animal that it is cleaning. It is a not uncommon sight to see two lemurs together, their heads in opposite directions, each cleaning the other's back.

Tarsioidea. The grooming activities of *Tarsius* are different from those of the lemur, owing perhaps to the absence of procumbent lower incisors and a sub-lingua modified as a brush. Le Gros Clark (1924*a*) has watched a *Tarsius* performing its toilet, and he describes it as being similar to that of a cat, the animal licking its fur. The hind-limbs, when it comes to their turn, are held outstretched with the hand of the same side. Two of the digits of the hind-foot are differentiated as scratchers, the other toes being held flexed when the animal is engaged in scratching. Mutual grooming has not been recorded.

Pithecoidea. The grooming activities of the members of this sub-order are altogether different from those of the lemurs and *Tarsius*. Amongst them, fingers are pre-eminently the implements of the toilet.

One hand is used to brush the hair aside, while the fingers of the other pick off particles from the skin and foreign matter adhering to the hairs. The tongue may be used to wet the hair, and odds and ends may be nibbled directly.

The fundamental pattern of grooming behaviour is apparently the same among all members of the sub-order Pithecoidea, and the sight of hair seems able to stimulate grooming during most phases of the life of a monkey or ape.* Different species differ, however, in the frequency with which they groom (thus gibbons are less conscientious about their cleanliness than are baboons) and in the extent to which lip movements accompany the activity. Although in all species grooming is more usually a performance in which two animals meet, a solitary animal may clean itself, or, on the other hand, as many as four may make up a single toilet party.

The possibility that fur picking in monkeys may constitute an act of adornment as well as a toilet has recently been raised by Tinklepaugh (1931), who describes an instance of a male monkey (*M. mulatta*) pulling out the hairs of the face of his female companion (a *M. irus*). "During this hair-pulling activity," writes Tinklepaugh, "the male frequently leaned back and surveyed the female as if evaluating the results." More data are necessary before this interpretation can be accepted as the only one that can fit the facts.

Nursing Behaviour

The manner in which nursing Primates carry their young is characteristic of each of the larger divisions

* Yerkes (1933) has suggested that the grooming behaviour pattern in chimpanzees increases in intensity with maturity.

of the order. It is sometimes possible to recognize familial distinctions.

Lemuroidea. Members of the genus *Lemur* carry their newly born young horizontally across the lower abdomen. (See Sclater, 1885.) The young animal holds tightly to its mother's fur, with both hands and feet, and its tail passes round her loins. The mother, as Mitchell has pointed out (1912), may sometimes help to support the baby with her own tail, "which she usually curls up between her legs over the body of the infant and then twists round her own body." At a later stage, when the young animal is fairly independent (? from about three months), it may be carried on its mother's back. As Wood-Jones has pointed out, however (1929), lemurs "do not nurse or handle or carry their young ones." Indeed, unless the mother-young relationship among the Lemuroidea is far more subtle than it appears objectively, it would seem that the young animal survives mainly because of its innately adjusted responses to stimuli presented by its mother's body. This cannot, however, be the whole story, but the extent to which the physiological condition of a recently parturient lemur forces her to show maternal responses still remains to be investigated.

Male lemurs apparently take little or no interest in their offspring.

The usual position of the young animal on the mother's body seems to be different in those families of Lemuroidea (e.g. *Lorisidæ*) in which there are inguinal as well as pectoral nipples. Wood-Jones (1929) provides the following description of the slow loris (*Nycticebus*):

> "In the case of *Nycticebus* the very dense woolly fur of the mother, and the small size of the newly-born young, renders

the concealment of the offspring very perfect. The difficulty of observation is here increased by the fact that this animal always bites savagely when interfered with and the coloration of the young is identical with that of the mother. So difficult is the little creature to detect that I have purchased, from a particularly shrewd native dealer, a mother and its offspring for the price of one animal, the presence of the baby having remained unsuspected. With this species the young does not cling transversely, but buries itself in the fur of the mother's ventral surface, with its hind limbs encircling the upper part and its forelimbs the lower part of the abdomen. The head is turned towards the mother's inguinal region of one side and the false nipple is grasped with the mouth. In this way the position of the young is inverted, its head being directed towards its mother's hind end during rest. From this position the young completely reverses itself when it wishes to suckle at the pectoral mammæ."

Tarsioidea. There are two reliable records about the nursing behaviour of *Tarsius*, those of Cuming (1838) and Le Gros Clark (1924*a*). The former writer had the good fortune to buy an adult female which some time later gave birth to a single baby. He writes:

"The young appeared to be rather weak, but a perfect resemblance to its parent; the eyes were open and covered with hair; it soon gathered strength, and was constantly sucking betwixt its parent's legs, and so well covered by its mother, that I seldom could see anything of it but its tail; on the second day it began to creep about the cage with apparent strength, and even climb up to the top by the rods of which the cage was composed. Upon persons wishing to see the young one when covered over by the mother, we had to disturb her, upon which the dam would take the young one in its mouth, in the same manner as a cat, and carry it about for some time; several times I saw her when not disturbed trying to get out of the cage, with the young one in her mouth as before."

Le Gros Clark failed to corroborate the observation that the young *Tarsius* is carried in its mother's mouth, but observed that it "always held on to the fur of its mother's belly," without being actively handled by her.

It is not known whether or not the male *Tarsius* pays any attention to its offspring. Le Gros Clark's observation (1924*a*) that *Tarsius* is always found in pairs except at the end of the breeding season, when the female and young are found alone, is not consonant with the established fact that *Tarsius* breeds at all times.

Pithecoidea. From the above descriptions it is plain that nursing behaviour in *Tarsius* is little different from that in the Lemuroidea. Both, however, differ greatly from the corresponding behaviour of members of the Pithecoidea, and within this sub-order the Platyrrhines differ from the Catarrhines.

Amongst the Platyrrhines the baby appears to be characteristically carried on the back, its head towards the mother's head and its feet and hands buried in her fur. The mother at first rarely, if ever, handles her baby, and the young one moves of its own accord to her nipples, which in many species are closer to the axillæ than is usual in catarrhine monkeys. When there are two young, as commonly occurs among the *Hapalidæ* (genera *Hapale*, *Leontocebus*, etc.), both are moored to the back.*

Among the members of the latter family the young

* Dr. C. R. Carpenter, who is studying the howler monkey (*Alouatta palliata inconsonans*) in Barro Colarado Island, informs me in a personal communication that "shortly after birth and for a period of approximately two weeks, a period which varies considerably with individuals, howler infants cling to the abdomen and lower ventral thoracic region of the mother. After this period, infants of this species are to be observed less and less frequently on the ventral surfaces of the mother, and more and more often on the sacral and lower lumbar regions of the back." Thus, except for a very short period following birth, the young of this species of New World monkey are carried, like young capuchins, on the mother's back.

may be carried from immediately after birth by the father and even by the offspring of the previous birth, if it is still with its parents (Lucas, Hume, and Smith, 1927). The mother, in the case described by these authors, "only received the baby in order to suckle it, which she did about every 2–3 hours throughout the day. She would approach the father and hold out her arm, when the baby would scramble, apparently spontaneously, on to her. After the feeding was over she would pull the baby off and give it to the father, who would take it and tuck it back on his flank." The young are also carried by the males in certain genera of the *Cebidæ* (e.g. *Saimiri*, see Zuckerman, 1932*a*, p. 191). The significance of this curious behaviour has not yet been investigated.

The father of a young capuchin at present living in the London Gardens (see table V, No. 9) at first paid no attention to his offspring, who was always carried on the female's back. The latter behaviour, judging from the naturalistic literature, is apparently characteristic of the genus *Cebus* (see Zuckerman, 1932*a*, p. 191). From the time when the baby was two months old and moved about its cage freely, it began to play with its father, who occasionally groomed it. Overtly this relationship was no different from that of any two unrelated animals.

In the catarrhine section of the Pithecoidea the baby usually hangs on to the belly of its mother, its feet grasping the fur of her belly and its hands buried in that of her chest. When the mother is sitting, her arm may sometimes be round the body of her offspring, whose legs may be on the ground. When the mother moves the young animal hangs on to her ventral surface, usually unsupported. When leaping or moving quickly, however, she may support her baby with one hand, but in some species this appears to be unusual. Thus the young hybrid mangabey with deformed limbs, noted in table V, No. 7, hung to its mother's belly only with its arms, its paralysed hindquarters dangling as its mother jumped about her cage.

At a later stage (circa one month) the young catarrhine monkey begins to leave its mother's body and to move independently in her neighbourhood. During this phase of its growth it will either return of its own accord to its mother, or will be snatched up by her when she moves, or when danger threatens. The mother also appears to control her offspring's first walking movements, and she grooms it assiduously.

At a still later stage (e.g. circa six months) the young animal may be carried on its mother's back (especially in baboons). From about this period other members of the group in which it lives pay attention to it, and carry it—both ventrally and dorsally.

It seems as though catarrhine males pay no specific attention to their offspring.

The mother-young relationship in apes is fundamentally the same as in catarrhine monkeys, although there appears to be greater maternal care and interest. Nevertheless, as with monkeys, the young ape is given little help in finding the breast by its mother. (For further information on this subject see von Allesch, 1921; Bingham, 1927; Fox, 1929; Nissen, 1931; Aulmann, 1932. Jacobsen, Jacobsen, and Yoshioka, 1932, report on the first year's growth of a chimpanzee whose mother died soon after its birth, and present data regarding the development of its various behaviour patterns.)

[Photo. by F. W. Bond

11. DRILL. Genus *Mandrillus*

[Photo. by F. W. Bond

12. CHACMA BABOON. Genus *Papio*

Other Behaviour Patterns of Classificatory Value

The above examples of behaviour patterns demonstrate clearly the differentiation of the main groups of Primates. Many more examples, typifying subordinal, familial, generic and specific distinction might be cited, but their definition and discussion cannot be attempted here. There is, for example, the "presenting" reaction exhibited by all catarrhine Primates that have been observed—the turning of the perineal region of the body of one animal into the line of vision of another. This form of behaviour develops very early in life, and is a conspicuous feature in the social lives of the Old World Primates (for detailed discussion see Zuckerman, 1932*a*). On the other hand, I have never seen it exhibited by either platyrrhine Primates or lemurs. Then there is the question of the vocalization of the different primate groups, and such characteristic specific behaviour as the chest beating of the gorilla. Facts like the latter clearly require examination from the point of view of their ontogeny and physiological or psychological background.* Nevertheless, even without such study, their sharp limitation to single species or to wider groups of Primates is a clear indication of their phylogenetic significance. There can be no doubt that an extensive study of the behaviour patterns of Primates would well repay the trouble entailed. At least some of them would prove to be of great interest from the taxonomic point of view.

* Yerkes and Yerkes (1929) have reviewed the relevant literature, and have concluded that chest beating is a specific form of behaviour of the gorilla, stating that it has only once been reported in the chimpanzee. To this instance can be added another, which I have reported elsewhere (Zuckerman, 1932*a*, p. 283). Both the fact that chest beating is characteristic of the gorilla, and that it is occasionally exhibited by the chimpanzee, are of interest from taxonomic and phylogenetic points of view.

CHAPTER VIII

THE DISEASES AND PARASITES OF THE PRIMATES

THE enthusiasm with which an old problem is assailed along a new path does not provide any criterion of validity either for the method of attack or for the results it obtains. It may also obscure an unbiassed view of the new approach. This in many ways is the position in which phylogenetic speculations based on comparative host-parasite studies stand to-day. The fame of this phylogenetic method has grown to such proportions that it cannot be passed over. But although the method undoubtedly has a real use, most of its exponents either fail to realize, or to point out, that it is a method which can never rank with the more conservative instruments of taxonomy. An animal's anatomy is the animal. So is its physiology, and its behaviour. The taxonomic relationships they reveal are definitely valid within the system of logic in which taxonomy operates. If the characters of a group of animal's parasites are in accordance with these relationships—well and good, and these characters may legitimately be explained in relation to some scheme of host-parasite evolution. On the other hand, the relationships revealed by direct study of the animals are not necessarily wrong if parasites fail to provide them with any support. In such a contingency some other explanation would have to be sought to explain the distribution of the parasites. For example, many human parasites show striking resemblances to those of other primate animals. Many do not, and are more closely related to the forms inhabiting domestic animals. Clearly our ideas about which of these parasites are significant phyletically will accord with predetermined views of man's zoological relationships.

The method, too, so far as the Primates are concerned, is limited in scope. This can be seen by turning to some of the evidence (Ewing's work on the genus *Pediculus*) which Metcalf (1929) adduced in support of his exposition of the host-parasite method in his paper on "Parasites and the aid they give in problems of taxonomy, geographical distribution, and paleogeography". Ewing (1927) points out that there are two distinct groups of *Pediculus* in the New World, the one confined to man, the other to monkeys. This fact he does not wish to be interpreted as indicating that if the lice and their hosts were traced back into the past, a louse ancestral to the two groups of *Pediculus* would be found inhabiting a Primate who would prove to be a common ancestor to man and the South American monkeys. On the contrary, he is at pains to suggest that the monkeys may at some time have been infected with lice by primitive man, and that the parasites have diverged from the human type in their new environment. This interpretation, if it is correct, does indeed suggest that a louse was equally at home on the monkey and on man, and that the two hosts were therefore in some way related. But this we already knew from sources far more incontrovertible. In this case, therefore, the lice provide primate systematics with doubtful support, at the same time failing to add anything new. Nevertheless, the host-parasite phylogenetic method cannot be denied exercise among the Primates simply because South American lice fail to do it justice. Even though none are known up to the present, it is possible that there exist parasites which will help to unravel the history of this order of mammals as much as the Opalinids are stated to have helped to unravel that of the frogs. In any case the facts about the parasites of Primates are of interest here even if they only relate slightly to the problem of the evolution of this mammalian order.

The host-parasite taxonomic method—which Metcalf calls "the von Ihering method"—can be extended to include non-parasitic diseases as well.* The many aspects there are to this wider approach to the problem of the relationships of the Primates were well demonstrated in the discussion on "Monkeys and Human Disease", held on February 7, 1929, at the Royal Society of Medicine, London (Pocock and others, 1929), when contributions were given from the angles of bacterial diseases, filterable viruses, protozoal parasites, helminth parasites, phthiriasis, and morbid anatomy. Obviously I cannot hope to deal here with more than a fraction of the available material. For admirable summaries of the facts, reference must be made to the report of the Royal Society of Medicine discussion, to Stiles, Hassall and Nolan's Key-Catalogue

* Metcalf calls this method the "von Ihering method", because it was von Ihering who about 1880 first emphasized the importance of host-parasite investigations in studies of evolution. In a book of anecdotes about the behaviour of monkeys, published anonymously by Stewart Rose in 1825, is a story of a monkey that died of some gastro-intestinal disorder on board a ship. In its illness it was attended by a surgeon who "pronounced that Pug died of the *iliac passion*, and announced this as a reason for believing that man was but a better breed of monkey." This seems a fairly good forecast of the host-parasite-disease phylogenetic method.

of Primates for which parasites are reported (1929), to Wenyon's work on protozoa (1926), and to Hegner's paper (1928) on the evolutionary significance of the protozoan parasites of monkeys and man.

The data to which I mainly intend to refer are those which throw some definite light upon the taxonomic relationships of the Primates—in other words, observations which refer to the segregation of parasites to definite groups of monkeys or apes. Statements of a more general character implying that these animals suffer from diseases similar to those of man, and that they are either the only animals susceptible to certain human infections, or that they are more susceptible to some human infections than are the usual laboratory animals, are of great interest, but for several reasons they are very difficult to interpret correctly from the taxonomic point of view. Most observations of this kind are based upon cross-infection experiments, and their taxonomic interpretation depends quite as much on the experimental technique on which they are based as it does on questions of natural susceptibility or resistance. Thus at one time the yellow fever virus could not be transmitted to any mammals except monkeys. That is the position to-day with regard to the transmission of poliomyelitis. But, with the aid of a new technique, small rodents and cats have recently been successfully inoculated with the yellow fever virus, and by comparison, it is difficult to maintain any longer that the causative factor of poliomyelitis is necessarily confined in its effects only to Primates. A different technique might prove such an hypothesis to be wrong.

Gaps in medical knowledge also prevent a proper taxonomic interpretation of other very suggestive observations made in medical experiments. For example, usual methods of inoculation, or experimental mosquito infection, prove adequate to transmit yellow fever, together with all its signs and symptoms, to members of the Asiatic genus *Macaca*, different species varying in their susceptibility. With similar methods, African Primates—species of *Cercopithecus*, *Papio*, as well as the chimpanzee—may be made to contract the disease in so far as the virus can be discovered in their bloods. As a rule, however, they fail to show any symptoms, the virus not proving pathogenic.

At first sight this difference between Asiatic and African Primates would appear to represent the geographical distribution of a natural immunity, since yellow fever is endemic in Africa and unknown in Asia. Such a simple interpretation of the facts, however, is not supported by the observation that monkeys indigenous to South America, where yellow fever is also endemic, are highly susceptible to the disease.* Before dismissing the question of susceptibility to yellow fever from this discussion, it may also be noted that experiments often show that there is a great difference in the virulence of the same strain of virus in man and monkey. A mild human infection may prove fatal in a monkey, and *vice versa* (see Hindle in Pocock and others, 1929). Much the same is shown, as is recorded below, by cross-infection experiments with malaria parasites.

Observations of similar interest have also been made

* For data relating to these questions, see Hindle's article shortly to appear in the Tropical Diseases Bulletin, 1933, the manuscript of which the author has very kindly allowed me to see. It may also be noted that Flexner and Lewis (1910) have suggested a difference in the susceptibility of New and Old World monkeys to poliomyelitis (on this point see also Jungeblut and Engle, 1932).

on other organisms—for example, *Treponema pallidum*, the causal organism of syphilis, and on a spirochæte of relapsing fever. Both of these organisms have been shown by experiments to thrive not only in man and other Primates, but also in non-primate mammals. Thus Clark and his co-workers (1931) found that the spirochæte harboured by a "squirrel monkey" (*Leontocebus*) in the Panama region was identical with the local species of human relapsing fever, and these workers readily passed the organism from the monkey, not only to other kinds of monkey, but also to human volunteers, white mice and rats and guinea-pigs. Here then is a disease without very definite boundaries in the mammalian world. The *Treponema* of syphilis is probably no more restricted in its possible range, and, as is well known, the signs it produces in the apes are far more like those it stimulates in man than are those it produces in monkeys (see Harrison, 1931).

It is easy to over-emphasize the phylogenetic significance of this last observation, if one ignores the fact that in other diseases man shares his misfortunes equally with animals far removed from his own zoological group. On the other hand, it does seem that the facts relating to the distribution of diseases such as those discussed in the preceding paragraphs are significant from the point of view of primate relationships, however little present knowledge allows them definite application in this field of study. At some future date it may be possible to apply them to such studies in the way that the distribution of certain ectoparasites, helminths, and protozoa can already be employed.

The Ectoparasites of the Primates *

Vermin are rarely found on monkeys and apes in captivity. Members of the Mallophagan genus *Trichodectes* occasionally occur on Primates both of the Old and the New World, but their distribution is not of such interest as that of the Anopluran family *Pediculidæ*. Lice belonging to this family are found on no mammals but the Primates. The genus *Pedicinus* is limited to members of the *Cercopithecidæ*, having already been found in the fur of species of *Presbytis*, *Colobus*, *Cercopithecus*, *Erythrocebus*, *Macaca*, *Cynopithecus*, and *Papio*, in fact in all the main genera of this family. The many specific differences in the lice affecting the different species and genera of this group of monkeys do not seem to be of interest in the taxonomy of the Primates. The same louse may inhabit the fur of monkeys geographically as far removed from each other as the Mona cercopitheque of West Africa and a macaque of India, while on the other hand two species of lice may inhabit the same species of monkey.

A few lice belonging to other genera of the *Pediculidæ* (e.g. *Neopedicinus*, *Phthirpedicinus*) are also found on Old World monkeys.

No lice of the genus *Pedicinus* have yet been found on any of the bigger apes or on the gibbons. Like those of man, the characteristic vermin of these animals belong to the genus *Pediculus*. Species of this genus have been found on gibbons and chimpanzees—and also on South American spider monkeys (various species

* The facts recorded below have been abstracted mainly from Stiles, Hassall and Nolan's Key-Catalogue (1929). Reference has also been made to Kellogg (1913, 1914), and to Ewing (1927). My attention was very kindly directed to the work of Kellogg and Ewing by Mr. V. B. Wigglesworth of the London School of Tropical Medicine and Hygiene.

[*Photo. by F. W. Bond*

13. GELADA BABOON. Genus *Theropithecus*

[*Photo. by F. W. Bond*

14. BLACK APE OF CELEBES. Genus *Cynopithecus*

of *Ateles*), on the brown capuchin (*Cebus fatuellis*), and on marmosets (*Leontocebus nigricollis*). Ewing has established the fact that *Ateles* is a true host for certain species of *Pediculus*, but authorities apparently regard the *Pediculus* lice that were found on the marmosets and capuchin as "stragglers". I have already noted Ewing's explanation of the occurrence of *Pediculus* on spider monkeys. Ewing also suggests that the lice of these animals should be referred to the sub-genus *Parapediculus*, and those of man to a sub-genus *Pediculus*.

Another interesting genus of *Pediculidæ* is the crab-louse, *Phthirus*, which hitherto has been found only on man and on the Kivu gorilla, *Gorilla beringei*.

It is almost unnecessary to define the obvious bearings of the distribution of all these ectoparasites on the phylogeny of the Primates. A few of the more salient ones may, however, be noted. In the first place it is seen that a single and distinct family of sucking lice is confined to one order of mammals—the Primates—appearing in all its major sub-groups. Secondly, one genus of this family, *Pedicinus*, is confined to one family of Primates—the *Cercopithecidæ*. Thirdly, it is possible to recognize a close resemblance between the lice of man and those of the apes.

The Helminths of the Primates *

Cameron,† who has made a wide study of the helminths of mammals, has often pointed out that in spite of fairly extensive investigations, comparatively little is yet known of the helminthic parasites of monkeys and

* I am indebted to Professor R. T. Leiper, F.R.S., of the London School of Tropical Medicine and Hygiene, for references to the literature of this subject. Stiles, Hassall and Nolan's Catalogue has again been freely used.

† In Pocock and others, 1929.

apes. Because of this, great caution is needed in interpreting from the point of view of primate relationships the information that is already available. Moreover, there are very few genera of helminthic parasites which appear to be restricted to the Primates, and differently segregated among its sub-groups, in such a way as to prove of use in primate taxonomy. Thus the genus *Ternidens* of the Strongyloidea, which is said (Baylis, 1923) to be limited in range to this order of mammals, has what may justifiably be described as a chaotic taxonomic distribution within the group (see Stiles, Hassall and Nolan, 1929), and this distribution quite plainly allows of no inferences of any sort regarding the relationships of this order of mammals.

Cameron (1926, and also in Pocock and others, 1929) has tried to classify human helminthic parasites on the basis of their probable history, but only one of the groups which he defines is of interest from the point of view of the evolution of the Primates. This group consists of the pin-worms. *Enterobius vermicularis*, which occurs in man, finds its closest relations in the pin-worms of the monkeys, and Cameron suggests that the human worm was a simian parasite "which first attacked man in his pre-human days." In an extensive study published in 1929, he showed how one *species* of *Enterobius* is restricted to one *genus* of monkeys—a fact which suggests that the evolution of the parasite is slower than that of the host. His further conclusions clearly support the accepted classification of the Primates—at the same time indicating the need for cautious interpretation. I quote them here in full:

"If one assume the existence of a *pre-enterobius* form in the *pre-simian* host, then the modifications of the parasite should accompany the generic differences of the host. One would expect to find forms most closely related to the human parasite in apes, while those in old world monkeys would be closer to *E. vermicularis* than those in new world monkeys and the lories, but not so close as in apes. This actually does seem to be the case although many species are inadequately described and many other species of monkeys have to be examined before the series will be sufficiently extensive to justify any results of value to anthropology."

Subulura, which Cameron described in 1930, is another genus of helminths of some interest from the point of view of the classification of the Primates. The worms included in this group are almost wholly restricted to birds, but six species occur in Primates. Three of these (from *Loris*, from *Tarsius*, and from *Galago*) are but little known. The fourth is found in various species of the genera *Cercopithecus*, *Erythrocebus*, and *Cercocebus*, and the fifth is distributed amongst genera of the *Cebidæ* and *Hapalidæ*. The sixth species is of great interest. It was created by Cameron, who gave it the name *S. distans*, to include peculiar forms of *Subulura* which he found in the wild green monkeys, *Cercopithecus sabæus*, that live in the island of St. Kitts in the West Indies. It is conceivable that *S. distans* has differentiated during the time that has elapsed since the monkeys were introduced to the island from West Africa.

The Protozoa of the Primates *

The order in which protozoa are discussed below is the same as that adopted by Hegner (1928) and by Thomson (in Pocock and others, 1929).

Amœbæ. Dobell (1931; references to other papers are

* I am greatly indebted to Mr. C. Dobell, F.R.S., for information relating to this subject.

given in this monograph) has shown conclusively that the *coli*- and *histolytica*-like *Entamœba* living naturally in certain macaques are specifically identical with those of man, and there seems little reason to suppose that the corresponding organisms which inhabit the intestinal tracts of gorillas, chimpanzees, and other Old and New World primate forms will prove to be in any way different. Thus *Entamœba* can provide no help in primate taxonomy. At present, moreover, it does not appear likely that any other amœbæ can.

Intestinal Flagellates. The genera *Trichomonas*, *Chilomastix*, *Giardia*, and *Embadomonas* occur in both man and many sub-human Primates, including apes and monkeys, the human and sub-human forms being indistinguishable morphologically.

Intestinal Ciliates. The balantidia of man are identical with those of the many sub-human Primates in which they occur—and possibly, too, with those of pigs and guinea-pigs. These organisms obviously have no significance from the point of view of primate relationships. There are some intestinal ciliates, however, which are of interest from this point of view. These are the two species of *Troglodytella* of the order *Oligotrichida*, which are found only in the gorilla and chimpanzee (see Wenyon, 1926; Swezey, 1932). This genus is allied to the *Entodinium* and *Diplodinium* of non-primate mammals, and in the absence of experimental knowledge it is difficult to assess the significance of its restricted primate range. This distribution may signify either similarity of the intestinal environments of the two African apes, or natural cross-infection of the two hosts, or their phylogenetic connection.

Malarial Parasites. The distribution and behaviour of these organisms are of great interest, but in the present state of our knowledge it is impossible to assess their true significance to the problem of primate relationships. Malarial parasites, many of which closely resemble the three forms found in man, occur in the apes and in several Old and New World monkeys (see, for example, Clark, 1930). They are also believed to cause the deaths of many monkeys, especially in the early stages of life (Clark, 1931). Cross-infection experiments (such as those of Clark and Dunn, 1931) have as a rule shown rather a rigid host-parasite specificity, and until recently infections had not been transmitted from man to any sub-human Primate (or *vice versa*), or with any success from apes to monkeys, or from monkeys of the New World to those of the Old. Recently, however, Knowles and Das Gupta (1932) have succeeded in producing a remarkable series of cross-infections, beginning with an original host *Cercopithecus pygerythrus*,* and passing the plasmodium through *Macaca mulatta* (= *M. rhesus*), *Macaca radiata*, *Macaca irus* (= *M. cynomolgos*), *Presbytis entellus* (= *Semnopithecus entellus*), *Hylobates hoolock*, to man. In its passage from one experimental host to another, the organism displayed amazing variations. In some its influence was altogether benign. In others (e.g. *M. mulatta*) it produced severe manifestations of malaria —and as a terminal event in some of the Rhesus monkeys, hæmoglobinuric fever. Its morphology, too, varied from host to host. In its original form in the African Vervet monkey it was similar to *Plasmodium vivax* of man; in the Rhesus monkeys it resembled *P. falciparum*; and in the human volunteers, *P. malariæ*.

* The original host of this series of cross-infections is elsewhere called "the Singapore monkey". This name hardly applies to *C. pygerythrus*, and belongs more properly to *M. irus*.

The appearance remained constant for each species, but back-crossing from the experimentally infected Rhesus monkeys to the original host species, the Vervet, did not stimulate the organisms to revert to their original *vivax* appearance.

As the authors of this report write, their results suggest an entirely new problem in protozoology, if not in parasitology generally. They might have added that the dynamic picture they have drawn of differences in reactions to the same protozoon are in striking contrast to the usual story of protozoal uniformity in the Primates. Apart from that of the ciliate genus *Troglodytella*, the distribution of these organisms in this order of mammals provides little or no aid to its taxonomy, even though the facts of protozoal similarities among the Primates may, as Hegner writes, be regarded "as evidence of importance in favour of the hypothesis that monkeys and man are of common descent."

CHAPTER IX

HYBRIDIZATION, AFFINITY, AND DIVERGENCE

IT is a fundamental assumption of biology that interbreeding implies close phyletic relationship of the animals that cross. How distant animals may be from each other phylogenetically, and yet successfully cross, is doubtful, and at this point arises one of the difficulties of taxonomy. The concept of a species as a group of individuals propagating their kind, and not effectively fertile except amongst themselves, has long been abandoned—rather it has never been truly applied—in practical systematics, for the simple reason that we do not possess information about the breeding habits of more than the minutest fraction of the total number of the different animal forms that are known. The acquisition of such knowledge is an experimental problem, and systematics and experimental zoology are different facets of biology. Yet systematists have by no means overthrown the Linnæan concept. On the contrary, it is still sometimes their custom, and that of their critics, to regard with suspicion the definition of a species which subsequently is found to disobey the Linnæan concept, and its implication that a species should be unable to cross successfully with a neighbouring but different one. Evolutionary discussion, too, is often greatly concerned with the idea of inter-specific barriers, in the sense of physiological and morphological impediments to effective hybridization.

As is now well known, however, absolute sterility and absolute fertility grade into each other in fairly definite stages. Robson (1928) has summarized the possible results of attempting to cross animal forms as follows:

"(1) Positive hostility between the male and the female.
(2) Absence of hostility {inability / disinclination} to pair.
(3) Coitus without reproduction.
(4) F_1 obtained, but with disturbed sex-ratio, ill-health or other abnormality.
(5) F_1 normal and healthy, but sterile.
(6) F_2 obtained, but weak or unhealthy.
(7) Fertile and viable F_2."

This summary suggests that inter-specific barriers need not only be morphological and physiological, but that they may also be of a kind conveniently termed psychological. In evolutionary discussion, of course, geographical or ecological isolation is also a very important factor. Here, however, this aspect of the problem is not being considered.

In its actual manifestations, the range from absolute sterility to absolute fertility does not by any means necessarily conform with a view that species more alike should cross more readily than species less alike. In this connection one may note the following examples, cited by Robson, that contradict such an hypothesis: (*a*) ordinal crosses amongst teleost fishes are sometimes more successful than crosses affecting families or genera; (*b*) two species of moth (*Pœcilopsis pomonaria* and *P. isabellæ*), which are conspicuously different morphologically, give fertile hybrids, but two species of fruit fly (*Drosophila melanogaster* and *D. simulans*),

[Photo. by F. W. Bond

15. PIG-TAILED MACAQUE. Genus *Macaca*

[Photo. by F. W. Bond

16. WANDEROO MACAQUE, OR LION-TAILED MONKEY
Genus *Macaca*

which are very similar structurally, give infertile hybrids; (*c*) a male moth caged with two females paired not with the one of its own species but with the one of a different species.

In spite of these occurrences, Robson still believes that there is a "general parallelism . . . between the degrees of mutual fertility and isolation . . . between morphological differentiation and reproductive capacity . . . between physiological affinity and morphological status." This parallelism is believed to be especially pronounced amongst mammalia, and inter-specific, and particularly inter-generic mammalian crosses, are usually regarded as matters of note. Thus the recently reported cross between a cow and an eland (Mitchell, 1932) received extensive notice in the Press. The matter was discussed in *Nature* by Haldane (1932*a*), who, referring to a previous paper of his, stated that he had been able to collect no more than six cases of inter-generic mammalian crosses, almost all concerning the family *Bovidæ*. Actually, many more have been recorded than those that Haldane notes (for example, all the inter-generic primate crosses referred to below), but his statement gives a clear idea of the prevalent opinion on this subject.

Scattered through a wide literature are occasional notes about the cross-breeding of monkeys. Although these references cannot be said to be records of controlled breeding experiments, they nevertheless appear to be reliable, for the reason that they are the records of the births of young to pairs of specifically or generically dissimilar animals confined alone in cages. In none of the published records to which I refer below, in tables IV and V, is there any hint of doubt regarding the authenticity of the crosses which are reported.

TABLE IV

Inter-Generic Crosses in the Family Cercopithecidæ

1. *Macaca irus* ♂ × *Cercocebus fuliginosus* ♀: the mother died near term 2/10/1878.

This cross occurred in the Zoological Gardens, London. (Sclater, 1878.)

2. *Macaca irus* ♂ × *Mandrillus sphinx* ♀: 14/10/78.

This cross also occurred in the Zoological Gardens, London. The sex of the offspring (which lived) is not given, but Sclater (1878) notes that it had a short tail.

3. *Macaca radiata* ♂ × *Cercopithecus pygerythrus* ♀:

The male offspring of this cross had the flesh-coloured face of a macaque and dark brown hair. (Gunning, 1909.)

4. *Macaca nemestrina* ♂ × *Mandrillus leucophæus* ♀: (Knottnerus-Meyer, 1904).

5. *Macaca maurus* ♂ × *Mandrillus leucophæus* ♀: (Knottnerus-Meyer, 1904).

The female of this cross was that of cross number four.

6. *Macaca irus* ♂ × *Mandrillus leucophæus* ♀:

The same female of crosses 4 and 5 was party to this one. The record again is that of Knottnerus-Meyer (1904), who remarks that the offspring of crosses 5 and 6 died soon after birth, probably having been born prematurely. The note does not signify clearly whether it also applies to cross 4. No mention is made of the sexes of the hybrids.

7. *Macaca nemestrina* ♂ × *Papio porcarius* ♀:

Sex of the offspring not given. It is stated that the hybrid resembled the young of *M. nemestrina*, but had the considerably more developed tail of *P. porcarius*. (Blyth, 1863.)

8. *Macaca irus* × *Papio cynocephalus* (*C. babuin*): (Schœpff, 1871, quoted from Ackermann 1898 and Przibram 1910).

9. This is a doubtful case.

Cercopithecus sabæus ♂ × *Macaca mulatta* ♀: 22/3/1873.

There was living offspring to this cross, which occurred in the Zoological Gardens, London, but there is some doubt as to paternity. In the Society's occurrence sheet for that day appears the note, hybrid "supposed to be between *Cercopithecus callitrichus* M. and *Macacus erythræus* F."

TABLE V

Inter-Specific Crosses of Primates *

Sub-order Pithecoidea

Div. Catarrhini

1. *Macaca nemestrina* ♂ and ♀ × *Macaca irus* ♂ and ♀: (Two instances: Gentry, 1872, Zool. Soc. Records (see Zuckerman, 1931)).

2. *Presbytis phayrei* ♀ × *Presbytis cristatus* ♂: (Sànyàl, 1893).

3. *Macaca mulatta* ♂ and ♀ × *Macaca irus* ♂ and ♀:

Numerous cases. Offspring fertile and of both sexes. (See, for example, Niemayer, 1868, who states that the young at first resemble those of either species, but later assume the shorter snout and shorter tail of the Rhesus. Knottnerus-Meyer, 1904, and Zool. Soc. Records (see Zuckerman, 1931)).

4. *Macaca mulatta* × *Macaca radiata:* (Zool. Soc. Records, see Zuckerman, 1931).

5. *Macaca radiata* ♂ × *Macaca irus* ♀: (Fitzinger, 1864; Rörig, 1903).

The crown hair of the female offspring of this cross was similar to the father's.

6. *Papio hamadryas* × *Papio cynocephalus:* (*C. babuin*)*:* (Isis, 1883, quoted from Przibram, 1910).

7. *Cercocebus torquatus* ♀ × *Cercocebus fuliginosus* ♂.

This cross recently occurred (10/6/1932) in the Zoological Gardens, London. The father is an albino, and the ♂ baby, which was killed when six months old (2/12/1932), had the coat colour of its mother's species. It was killed on account of deformities resulting from some post-natal injuries. Its eyes were light blue when it was born, and deepened slightly during the time it lived. All other mangabeys I have seen, in fact almost all monkeys and apes, have unmistakable brown eyes. The eyes of such monkeys (i.e. macaques, capuchins) as I have observed soon after their births were bluish, but changed to an unconditional brown always within two months. The persistent blue of the hybrid mangabey's eyes is almost certainly related to the peculiarities of its parentage. The cross is discussed more fully below.

Sub-order Pithecoidea

Div. Platyrrhini

Family *Hapalidæ*.

8. *Hapale jacchus* ♂ × *Hapale melanura* ♀: (English, 1932).

---

* Since inter-generic crosses involving most of the species of this table have been recorded, the list given here has been curtailed by the exclusion of certain unimportant cases.

Family *Cebidæ*.

9. *Cebus fatuellis* ♀ × *Cebus albifrons* ♂:

This cross occurred recently (14/6/1932) in the London Zoological Gardens. The baby, a male, is still alive, and its coat colour immediately after birth was predominantly like that of its father. The hair, however, appeared coarser, and the ventral creamy colour of the chest, abdomen, shoulders and forehead was slightly darker. A central vertical bar of brown hair divided an otherwise more or less bare forehead. Its face was darker in colour than its father's, and more like that of its mother. When six months old its coat was almost a perfect blend of those of its parents, hair having grown on the bare temporal regions. By the time it was two and a half months old the young animal's eyes were hazel-brown and thinly rimmed with blue. This pattern is exactly similar to that of its father, the mother's eyes being a uniform rust-red.

Sub-order Lemuroidea

Many inter-specific crosses of the genus *Lemur* have been reported (see Knottnerus-Meyer, 1904; Pocock, 1911). I have reported elsewhere on eight hybrids of this genus which were born in the London Gardens (Zuckerman, 1932*d*).

* * *

Before discussing the phyletic significance of these crosses, certain practical points relating to the question of hybridization of Primates may be considered.

No member of the *Colobidæ* figures in the table of inter-generic crosses; this fact need not be taken to mean that it has been proved that these animals are incapable of crossing with other members of the *Cercopithecidæ*. For one thing, the records I have found may not be all that have been published; and for another, langurs are notoriously bad breeders in captivity, even in countries of their natural distribution. Only one breeder of this sub-family has ever lived in the London Zoological Gardens (1912–1915). Clearly, too, the records given above cannot be taken to represent the limit of the possible combinations and permutations of inter-generic and inter-specific crossing in monkeys. The births I have recorded simply represent the accidental results of confining, from an early age, animals of different genera in the same cage. Deliberate attempts to cross-breed Primates have not, to my knowledge, been reported.

Several conditions would have to be observed in experiments on the hybridization of monkeys. Thus although it appears that these animals breed at any time, conspicuous seasonal variations occur in their birth-rates (see Fig. 1, p. 30), and the maximum of fertility in one species might coincide with the minimum in another (e.g. *M. mulatta* and *P. hamadryas*). It is likely that success in hybridization would be better achieved by two species whose seasonal variations in fertility coincide, than by two in which they do not.

Another question which would have to be considered is the mechanical ability of animals of different genera or species to pair with each other. Incompatibility might be due either to discordant body-size, or to the inter-specific and inter-generic differences which are found in the form of the external genitalia of both sexes. These organs have been considered in great detail from the taxonomic point of view by Pocock (1917, etc.). In the female the variations mostly concern the sexual skin, for example the degree of swelling it undergoes, and since the external genitalia of a male are adapted to those of the female of its own species, it follows that the chances of a successful cross will be greater if males are mated with females which exhibit sexual-skin variations comparable with those shown by their own species. But as the record of such a cross as *Macaca irus* ♂ and *Mandrillus sphinx* ♀ indicates, discordant body-size and external genitalia are not a bar to successful mating.*

The hybridization of two species or genera might also be effectively prevented by differences in their gestation periods. This might conceivably be an explanation of the still-birth of case 1 in the table of inter-generic crosses. But as I have noted on p. 28, the duration of pregnancy is very much the same in the four Old World monkeys and the one New World form about which we already have information, and it does not therefore seem that this could be a serious limiting factor to cross-breeding in the family *Cercopithecidæ*.

Yet another factor must be considered in the cross-breeding of Primates. It is of a kind that would be described as psychological, and is one which I have discussed elsewhere (Zuckerman, 1932*a*) under the term "dominance". When two monkeys are introduced to each other it is usual, especially when there is some disparity in size and age, for the one to be dominated by the other. Such dominance might take a violent form, and, if the animals are of different species, it might superficially be regarded as the pugnacious hostility of one species to sexual contact with another. A wide study of this behaviour, however, gives little

* At the time of writing (10/12/32) four months have passed since an adult male *M. nemestrina* (pig-tailed macaque) was isolated with an adult female *P. anubis* (Anubis baboon). The male is the father of three young pig-tailed macaques born in the Gardens, and the female, although there is no record of her ever having borne young, is apparently in perfect health. Her menstrual cycles, which have been observed for almost four years, are very regular. Although the animals often mate, conception has not yet occurred.

support for such a view. Monkeys and apes, moreover, have repeatedly demonstrated a very catholic taste in their choice of sexual partners (see, for example, Hamilton, 1914). Nevertheless, although an adult male of one genus will sometimes immediately assume amicable relations with an adult female of another genus, it is probably best in hybridization experiments to bring young animals together, and to allow them to grow up in company.

* * *

It is perhaps justifiable to deny, on the score of imperfect systematic definition, any phyletic importance to some of the interspecific crosses that have been cited above. Such an attitude cannot be maintained about the inter-generic crosses. If the term genus has any phyletic significance at all, it clearly has this significance in its application to the more widely different monkeys—e.g. baboons, macaques and cercopitheques. These forms have been separated from one another at least since the closing phases of the Miocene, that is for some 10,000,000 years. The fact that such types can cross-breed is hardly indicative of the existence of physiological or morphological inter-generic barriers, and clearly shows that the amazing variety of gross morphological differentiation that has occurred in them since the Miocene has not gone hand in hand with a physiological and cytological differentiation. In spite of the almost complete absence of data relating to the fertility or sterility of the offspring of these crosses, these facts throw a new light on the problem of the divergence of the Old World Primates from a common ancestor.

If the occasional notes about the appearance of the hybrids (summarized in the above tables) are to be relied upon, it would seem that the generic and specific characters of monkeys (e.g. tail in length, hair pattern and colour, and length of snout) behave in hybridization like varietal differences.

I have referred in some detail to the appearance of the hybrids in crosses numbers 7 and 9 of table V. The main difference in the species of *Cebus* concerned in cross number 9 is the thicker, coarser and uniformly darker hair in *fatuellis* (the female). This animal also has two dark brown tufts on either side of the head. The deep brown hair reaches to the margin of the face, which is a purplish flesh colour. *Cebus albifrons* (the male) has no crests, and his forehead is pale and almost bare. Short creamy hair reaches to the back of the head from the region of the coronal suture. Cream-coloured fur also covers the throat, shoulders and ventral surface of the body. The rest of the body is darker in colour. The eye-colours of the parent animals have been referred to in table V. Apart from these superficial differences between the two species, no others of note are recorded in the systematic literature. It is probable that the territories of the two types overlap to some extent. In view of the superficial differences of the parent species and of the appearance of their offspring, which I have described in the table, it is likely that the specific difference between them would prove to be slight when analysed genetically.

The difference would also appear to be slight in the

case of the species *Cercocebus fuliginosus* and *Cercocebus torquatus* concerned in cross No. 7, table V. And here the fortunate circumstance of the albinism of the father allows one to see slightly more deeply into the nature of the divergence. The species *fuliginosus* is normally characterized by sooty black fur on the upper parts of the body and on the outer sides of the limbs. The cheek-whiskers, throat, underside of the body and inside of the limbs are ashy-grey. The middle line of the back is the darkest part of the body; there is a crown patch of black hair, each with a central olive-brown band of variable size. The eyes are brown.

Torquatus is much lighter in colour and has a cap of cream-coloured hairs, which, as in *fuliginosus*, are tipped for $\frac{1}{4}$–$\frac{1}{2}$ an inch with black.* The upper part of the body and outer side of the limbs are dove-grey to a brownish drab. A purplish tinge may sometimes be detected in this part of the coat. There is a line of darker hair running down the middle line of the back. The less exposed parts of the body are lighter in colour. The eyes are brown. There are no obvious differences between the species apart from coat colour, and Schwarz, in his recent review of the genus (1928*c*), regards them only as sub-species of a single species. The two types are close neighbours in their distribution, but they are not known to overlap.

The father, in cross No. 7, is an albino of the species *fuliginosus*. His fur is completely white, with no trace of pigmentation. His eyes, on the other hand, though pink, show a distinct trace of blue in the iris—something more than the opalescent cloudiness of the eyes

* This is the condition in the specimen under consideration. Schwarz (1928*c*) describes the crown patch in this species as blackish brown with a posterior creamy zone

of an albino rabbit. (He suffers from nystagmus and always shades his eyes when looking closely at an object, particularly in bright light, thus apparently diminishing the intensity of retinal stimulation.) It is probable that his albinism is due to a single gene mutation as in the case of all other albino mammals which have been studied genetically. The trace of blue pigment in the iris suggests that the albinism may not be quite complete.

The hybrid, a male, was like its mother in coat colour except for its occipital patch which consisted of brown gold hair tipped with black. Descriptions of the species *torquatus*, however, allow of variation in the colour of this area. The difference between mother and child may, therefore, have no special significance in the present case; it may be an age or sex difference, or there may be considerable individual variation within the species. The eyes of the hybrid were perfectly blue, entirely unlike the eyes of either parent, or of either parent species (see table V, No. 7).

Obviously it is impossible to draw any certain conclusion from this single cross, and it is hoped to discover more about the differences between the species *torquatus* and *fuliginosus* by crossing together normal individuals of the two species, as well as by controlled matings of the hybrids. Meanwhile one is tempted to speculate, and at least two possibilities present themselves, each dependent upon a single gene mutation. If the albino male *fuliginosus* owes his albinism simply to the absence of a colour developer, as is known to be the case in some albino mammals, then the colouration of the female species, *torquatus*, must be dominant to that of the male species, *fuliginosus*. If, however, the albino owes his albinism to the absence of the colour

[Photo. by F. W. Bond

17. SOOTY MANGABEY. Genus *Cercocebus*
The speckling on the coat is sawdust

[Photo. by F. W. Bond

18. PATAS MONKEY. Genus *Erythrocebus*

[*face page* 104

base, as is also the case in many male albinos, the only possible colour of the hybrid would be that of the mother and the question of dominance does not arise —either the *fuliginosus* or the *torquatus* colouration may prove to be dominant. The blue eyes of the hybrid may well be due to delay or failure of the development of the pigment of the iris stroma owing to the presence in the hybrid of only a single "dose" of an essential colour factor.

Either of these suggestions involves the supposition that the colour difference between the species *fuliginosus* and *torquatus* depends upon a single gene. It is of course possible that many more genes are involved, but the results at present available can be perfectly well explained on a single gene difference.

Interspecific differences of such an order would certainly be unusual. In this instance, however, it must be remembered that Schwarz regards the divergence as being only a sub-specific one, and that the genes in which the species differ may occasion other non-obvious physiological differences which help to keep the two types distinct. The distribution of the group * of mangabeys to which these two "species" belong is imperfectly known, and it is likely that geographical isolation helps to keep the "fuliginosus" and "torquatus" genotypes relatively pure. With a uniformly sooty, a white-capped, and a red-capped type, the group constitutes an admirable one for experimental genetical study.

The possibility that only a slight factorial difference separates two accredited primate species—(or two subspecies)—suggests that there may not be very wide chromosomal differences between other primate species. It also stimulates one to enquire whether the systematics

* Called by Schwarz *Cercocebus torquatus*.

of the Primates has been pursued less critically than the systematics of other orders. When one considers the real and distinct differences between almost all the species recognized in this order of mammals, however, it cannot be doubted that they are taxonomically at least as valid as are most other mammalian species.

It is, nevertheless, conceivable that the different genera of *Cercopithecidæ* are in fact less divergent from each other from the genetic point of view than are, for example, those of such a family as the *Canidæ*. It is impossible to judge of this by means of any common denominator between the families, and one can only speculate that this may be so from an impression gained by a study of the relative proportions of inter-generic crosses among Primates and other orders of mammals. The point is not worth labouring owing to its altogether speculative character, and more opportunity may have been given monkeys of different genera to cross than has been given other mammals of different genera, but it does seem that relative to the number of primate births that have occurred in captivity, a larger proportion are hybrids than is the case in other orders of mammals.

The few facts there are relating to the fertility of hybrid monkeys refer only to the F.1 generation of the cross between *Macaca mulatta* and *Macaca irus*. Such hybrids are said to breed freely. Unfortunately, monkeys grow and breed slowly compared with the more commonly used experimental animals, and I have been unable to trace the later histories of the animals recorded in tables IV and V. It should be remembered, too, that the *Cercopithecidæ* in Zoological Gardens have "an average life of under seven years" (Flower, 1931), and as they do not reach sexual maturity until they are about five years old, those that are born in captivity as a rule have little opportunity to display their reproductive potentialities.

Since cytological differences are said (see Haldane, 1932*b*) to be largely responsible for interspecific sterility, the fertility or sterility of hybrid monkeys may be considered from the point of view of cellular anatomy. According to Painter (1924, 1925), one of the foremost investigators of the chromosomes of mammals, the diploid number in the Rhesus macaque is, as in man, forty-eight; the sex chromosomes are similar in the two types, but although in both there is the same general seriation in the size of the autosomes, they differ somewhat in their form in the two primate species. These observations have recently been confirmed by Evans and Swezy (1929). In a capuchin (*Cebus fatuellis*) which Painter also studied, the diploid number was fifty-four. In this New World Primate there were many more small chromosomes than were found in the Old World Primates, but the sex chromosomes were similar in both groups.

In the well-known survey of the chromosomes which he published in 1925, Painter pointed out that in spermatogonial metaphase plates "the chromosomes of man, the rhesus monkey and the bat are so similar that one would have difficulty in distinguishing between these species." The greater number of chromosomes in the New World monkey he regards as due to the fragmentation of some of the larger elements of a 48 series, which he considers to be the primitive number for Eutherian mammals. Such a process would not in itself, so he states, exercise any influence on evolution.*

* Painter develops the thesis of chromosomal homology in his paper on the chromosomes of the rat and mouse (1928).

The similarity of the chromosomes in such widely separated forms as the Rhesus macaque and man allows one to expect at least as close a similarity between the macaque and its fellow members of the family *Cercopithecidæ*. It is thus perhaps permissible to speculate that the F.1 generation of inter-generic crosses in this family will very likely be found to be fertile, and indeed it is possible that the only obstacle to successful inter-fertility between different species and genera of monkeys is broken down when they have been successfully made to mate with one another. Certainly there cannot occur any process antagonistic to successful hybridization comparable with the disorganized maturation division in the gametogenesis of the mule.

The facts considered above immediately stimulate an enquiry into the nature of the several processes that must have been related to the evolutionary divergence of the Old World Primates. In the present state of knowledge it may be tentatively assumed, for purposes of discussion, that the evolution of the Old World Primates was due to selection and isolation operating upon gene mutations and recombinations of genes resulting from hybridization. The process does not seem to have depended on polyploidy—and one may remember here Painter's conclusion that this process could not have played an extensive rôle in the differentiation of any Eutherian mammals.

In the case of the family *Cercopithecidæ*, and very likely in other families of the Pithecoidea, isolation does not appear to have depended on the development of fundamental physiological barriers between the different genera and species. Actually it is difficult to understand why the development of such barriers should so often be emphasized as being one of the more important factors operating in the evolutionary divergence of vertebrate groups, and why, unless for historical reasons, this criterion of species isolation is so often insisted upon to a disproportionate extent in present-day discussions of vertebrate evolution. As a hang-over from a period in which species were regarded as divinely created groups of animals breeding purely and only amongst themselves, it would be easy to understand. It is also comprehensible as a facile explanation, in the early days of evolutionary discussion, of the fact that species continue through the centuries separate and apparently unchanged. But to-day when the fact that next to nothing is known about the breeding potentialities of any except the smallest fraction of the world's animal fauna is generally admitted, it seems unnecessary to insist, as Bateson (1928) did, that the "production of an indubitably sterile hybrid from completely fertile parents" is the "particular and essential bit of the theory of evolution which is concerned with the origin and nature of *species*." Such a mechanism of isolation, as Gates (1925) too, and many others, have stated, may have played an important part in evolution by segregating from each other organisms which might otherwise interbreed, but until more is known about the incidence of interspecific sterility, it is clearly unnecessary to emphasize this factor of isolation more than any other factor having the same effect. There is, indeed, a great deal of evidence suggesting that the Linnæan species interpreted as "a group limited not merely by the anatomical resemblance of its individual members but also by *their inability to breed successfully with other forms*" (Hogben, 1930)—is a concept incapable of wide generalization in the problem of the evolutionary divergence of vertebrates. The *Cercopithecidæ* demonstrate its limitations very clearly. Among them neither evolutionary divergence nor geographical isolation seems to have resulted in the production of inter-specific and inter-generic barriers in the sense of physiological or morphological impediments to interbreeding ; and in their case there can be no question about the taxonomic validity of the vast majority of the different genera and species which comprise the family.

It is possible that some biological value could be attached to the concept of a species as defined above, and that it could be put to some use in the classification of breeding units. But obviously such a definition would be more or less useless, at the least cumbersome, as a basis of a scheme of classification which could form a foundation for the study of phylogeny.

The factor of isolation nevertheless remains extremely important. If an animal reacts either to a single gene mutation or to several gene mutations, or to a recombination of genes, by developing a new observable character (or characters), the new character will be preserved and spread only (*a*) if it is simultaneously developed in a sufficient number of other individuals, or (*b*) if, though developed in one or only a few individuals, the character has some very strong selective advantage either in the environment in which it is developed or in some other accessible one, or (*c*) if, though present in very few individuals, a geographical or other form of effective isolation protects it from random extinction. So far as the history of the Primates is concerned, nothing is known about either (*a*) or (*b*). Enough is known of the habits of the Old World forms, however, to throw some light on the possible operation of geographical isolation in promoting the spread of a particular variant genotype developed in a population either by crossing or as a result of mutation.

From what is known of the social behaviour and the natural lives of these animals, it would seem incorrect to discuss inheritance among them on the assumption that they form random mating populations. Most Old World Primates—apart from man—appear to live in small, and seemingly well-segregated, parties, consisting either of separate families dominated by a single male, or of several families, each of which strictly maintains its identity within the group it helps to form. Sexual selection operates intensely—though not in the Darwinian sense, since the females do not appear to have any control over their own disposal—and only a very small proportion of males succeed in propagating their kind. These males retain hold of as many females as possible, and prevent other males from coming into sexual contact with them. This they may do either by isolating themselves and their females, and actively avoiding contact with their fellow males, or by virtue of personal force, which is not necessarily a matter of physical strength. Thus in certain circumstances isolation may mean effective topographical separation of the members forming a homogeneous group; alternatively, the families of such a group, though remaining independent, may still live in fairly close contact in a relatively small area. It is this relatively close contact which apparently results in the formation of the larger groups consisting of several families. To what extent these comprise permanent associations is unknown. Although baboons are apparently the most gregarious sub-human Primates, they often live in single families, and it is likely that such small parties constantly split off from the larger groups, to roam in solitude. (For data relating to these questions, see Zuckerman, 1932*a*.)

In this brief summary of what is known of the gross organization of the social lives of sub-human Old World Primates, I have tried to emphasize the relationship between the urge that exists in these animals to secure as many females as possible, and the urge to protect these females from the attentions of other males. Where there is great discordance in the degrees of dominance of the males within a group, it would seem likely that those males who possessed females would be able to hold them in spite of being in close association with their fellow males. On the other hand, if no such state of social balance existed, it is equally clear that males securing females would immediately seek fresh country, and so avoid sexual clashes. These suppositions are inferences from observations that have been made in the field, but they are also supported by observations made, both during times of peace and times of strife, of the colony of baboons kept in the London Zoological Gardens. Here we have a possible reason for the dispersal and isolation of small groups of Old World Primates. It is probably the chief reason, since there is some evidence that in equable circumstances each group of monkeys or apes will for years keep to the same area. Certain inferences, also relating to the problem of dispersal, can be drawn from these suggestive, although admittedly incomplete, observations. Firstly, one may assume that the smaller a group is and the greater the sexual competition within it, the greater would be the probability of its splitting into separate family parties. Secondly, it would also appear likely that in such a case the area of dispersal of the group would be progressively wider the better the external conditions, and the less the competition from other forms. Conversely, if the type of environment to which the animals were adapted was restricted geographically, the animals could not separate so widely and the chances of the development of daughter colonies with genotypes differing relatively constantly in any particulars would decrease—although the amount of variation in the whole group would probably be greater as a result of random crossing.

The picture I have tried to draw of the social interactions which to-day lead to the isolation of groups of monkeys can be applied to earlier stages in the divergence of the Old World Primates. Since there is no reason to believe the opposite, we may assume a monophyletic origin for this group of animals,* and unless the first monkeys behaved altogether differently from those of the present day—and there is no reason for assuming that they did—their further evolutionary divergence must have been assisted by migrations resulting from sexual competition. The world of forests lay open, and they could spread far and wide—as indeed there is evidence that they did (see Black, 1925)—thus establishing separate small colonies in which the amounts of their "potential variabilities" would be limited owing to the fact that the particular genotypes forming these colonies would not be re-absorbed into the general population. I am, of course, assuming a high variability for the original group, increasing from time to time through hybridization and through mutation, both of which processes would also affect the small daughter colonies. That such a process continues even to-day is suggested by the existence of distinct

* Monophyletic in the sense of derivation from a single ancestral type. The "derivation" may have occurred repeatedly.

geographical races, "taxonomic sub-species", of such animals as the Chacma and yellow baboons of Africa.

It is impossible to say whether events of such a kind, aided by natural selection, would be sufficient to account for the wide differences in the form, in the feeding habits and in the habitats, of the different primate genera. Here, however, one faces the same problem that is met with in regard to other vertebrate families.

In the earliest days of the divergence of the Primates, the chances of migrating groups of Old World Primates settling in a territory already occupied by other members of this group must have been slight. As their numbers increased, however, so these chances increased, and in time groups altogether different morphologically must have found themselves in the same area—as they find themselves to-day. The Gelada "baboon" and the Hamadryas baboon are stated to inhabit contiguous cliffs. The Chacma baboon and the Vervet monkey can be found living in the wooded slopes of the same hills. The question that has to be answered is why the groups do not interbreed in such circumstances, since there is suggestive evidence that they can, in fact, do so. This question raises a more general consideration.

Many writers who are careful to show that the problem of the origin of the new species (in the Linnæan sense) is not directly relevant, as a definite issue, to the wider problem of the origin of new varieties (and *ipso facto* of new genera, families, orders, etc.), nevertheless refer to the question of "species incompatibility" as a problem whose solution will remove from the evolutionary hypothesis any "ulterior differences in the way of explaining the origin of differences which separate the larger systematic groups" (Hogben, 1930).

[Photo. by F. W. Bond

19. PLUTO MONKEY. Genus *Cercopithecus*

[Photo. by F. W. Bond

20. MONA CERCOPITHEQUE. Genus *Cercopithecus*

That is to say, the term "species incompatibility" is used as though it actually defined a universal biological problem presented by facts of observation. But as I have already suggested, the idea of "species incompatibility" is, on the one hand, at least partly an unwarranted generalization based upon an interpretation of species as groups of individuals incapable of breeding successfully with other groups of animals. A generalization of such importance cannot stand just as an inference from the facts that different species generally do not mingle, that they often prey upon each other, and that like begets like. It could not be established except by careful and extensive experiments, and such experiments have certainly not yet been carried out in sufficient number to provide any generalization at all. The essential problem suggested by the facts is not that of species incompatibility but that of species cohesion, not a problem of why animals of different species are incompatible, but a problem of why similar animals remain together. This distinction can be readily exemplified.

From the seeming fact that different genera and species of the *Cercopithecidæ* can interbreed, this group of animals would appear superficially to be similar to the different strains of poultry or different breeds of dog, to which references are so often made in discussions of the problems of evolution. Each of these groups shows conspicuous variations in structure, which may sometimes prove a hindrance to the successful crossing of the members within each group. Yet the three groups are profoundly different from the point of view of their internal relationships. Given that there were no human interference in either case, it is fairly certain that the hens and the dogs would, in time, both form homogeneous groups. The monkeys, on the other hand, would, relatively speaking, remain as they are to-day. Nevertheless, there is no obvious reason why the physiological and morphological differences between the various members of these three groups should not turn out, on genetic analysis, to be very much of the same kind and order.

Here then is what is perhaps the most intriguing aspect of the problem of species barriers. Why will many types not interbreed with each other even though they can do so successfully? What, in other words, is the basis of species cohesion? A problem as wide as this would almost certainly have a somewhat different —and exceedingly complex—answer in each animal group, varying with the particular dynamic organization of each different type of animal.

Obviously, so far as most Primates are concerned, one reason why an animal of one kind mates with those most like itself is the fact that those most like it are also those who are geographically nearest to it. But this will not explain the apparent non-interbreeding of different species, even genera, which may live in the same square mile of country, or which may mingle from time to time. And this "species purity" is especially curious when the catholic sexual taste of monkeys and apes is remembered. In captivity they have been seen to use dogs, snakes, and even inanimate things as sexual objects. Of course the purity of primate species may be more apparent than real. Hybridization may actually be taking place constantly. Banks (1929) has provided suggestive evidence that types of monkey which taxonomically are regarded as distinct species may form a single interbreeding unit. The species he has studied, as was mentioned on p. 12, were *Presbytis chrysomelas*, which is found in Sarawak over the plains and low hills, rising to about three thousand feet on the higher mountains, *Presbytis rubicundus*, which is never seen on the plains and is mostly mountainous, being found as high up as six thousand feet, and *Presbytis cruciger*, more or less confined to the low hills and lower slopes of mountains. The coat colours of their species suggests that *P. cruciger* is a hybrid resulting from the crossing of the langur of the plains and the langur of the mountains. Alternatively it is possible that the lower two species form a single breeding unit.

But here we are dealing with closely allied species, which may well be expected to cross occasionally. A male who is ejected from one group owing to sexual competition may yet succeed in capturing females from a neighbouring one. But the rarity of reports of heterogeneous groups of monkeys being observed suggests that such hybridization, especially between the more dissimilar species and genera, is at best clandestine and unusual, and this belief is strengthened by the frequency of accounts of contests between different genera, species, and even groups of the same species, for such prizes as a particular territory.

In the absence of any definite knowledge about the nature of species cohesion in the Old World Primates one can but assume, as a working hypothesis and for want of a better term, that it consists largely in a process of the social conditioning of the young animals born into a group. I have already attempted to show (Zuckerman, 1932*a*, Chapters XVII and XVIII) how unorganized are the first social responses of sub-human Primates, and I have also tried to trace these diffuse reactions as they become adapted to the habit of the group into which the young are born. This development must inevitably involve adaption to the specific attributes of the group (visual, olfactory, etc.), through whatever processes constitute learning in these animals. Because of such adaption, for "social traditions" of such a kind, baboons seek the company of baboons, and cercopitheques that of cercopitheques; and in the absence of other baboons, or of other cercopitheques, they will seek—given a complete freedom of choice—those most like themselves. There can be little doubt that the more complex the organization of an animal from the point of view of its sensori-motor mechanisms, the more definite and at the same time the more adaptable could this process become.

The divergence of primate species and genera, and their subsequent maintenance, can thus be related to the processes of mutation and selection, interacting with tendencies to isolation inherent in the behaviour of sub-human Primates. These tendencies may be briefly described as (*a*) the urge for family parties to seek isolation so as to avoid sexual clashes, and (*b*) the "social conditioning" of the members of individual groups to the characters of their groups, whatever they happen to be.

This hypothesis is amenable at least to partial experimental test. It also suggests that in an earlier period in the history of the Old World Primates, when the morphological divergence between the forms comprising the group was less than it is to-day, hybridization, whether it was clandestine or otherwise, may have occurred more frequently owing to greater similarities between types, thus providing a greater variety of genotypes which could then either thrive and alter by mutation in the isolation that was stimulated by sexual competition, or disappear in the stress of natural selection.

CHAPTER X

THE PSYCHOLOGICAL MEASURE OF INTELLIGENCE IN PRIMATES

THE questions to be discussed in this and the following chapters are in many ways the most interesting of all that are raised by the comparative study of the order Primates. Man is the product of the evolution of a simple Eocene animal; his pre-eminent position in the world he owes mainly to his peculiarly intelligent behaviour. Unless his linear ancestors were always gifted with immensely greater capacities for adapting themselves to external conditions than are usual among other mammals, as well as with the ability to change these conditions, it follows that his present eminence is due either to natural evolutionary changes or to the sudden interpolation into his psychological make-up, by some supernatural power, of a force unrepresented elsewhere in the living world. As there is absolutely no reason for assuming that man's pre-human ancestors were more intelligent than were their fellow-animals, either evolutionary change or miraculous divine intervention lies at the back of human intelligence. The second of these possibilities does not lend itself to scientific examination. It may be the correct explanation, but from the scientific point of view it cannot legitimately be resorted to in answer to the problem of man's dominantly successful behaviour, until all possibilities of more objective explanation through morphological, physiological and psychological observation and experiment are exhausted. To-day we are only beginning the researches which may provide the materials for stating in definite terms possible evolutionary stages in the development of human intelligence. The meticulous morphological and physiological studies of the primate brain on which many students have been engaged during the past fifty years have provided a foundation, and the past few years have seen the beginnings of what promises to be a successful experimental attack on the problems of the mechanisms of intelligence and the evolution of cortical dominance. This evolution may be defined as the reference to the cortex, in forms that are more highly organized neurogically, of functions which in lower types are subserved by more primitive parts of the brain. In the following chapters an attempt is made to state very briefly the present position of certain important studies on the Primates relating to these questions.

The Experimental Study of Primate Behaviour

Every animal's behaviour is chiefly a function of its brain and its sensori-motor apparatus. A dog cannot do the things which a monkey can do in virtue of its possession of hands. It cannot do the things a monkey does as a result of having stereoscopic vision. Attempts to define these differences of animal behaviour—especially the differences between the behaviour of the sub-human Primates and the behaviour of other mammals—were made long before animal psychology had passed out of its anecdotal phase. They were of little, if any, scientific value. Thorndike's comparative studies of behaviour, which date from the end of the nineteenth and the beginning of the twentieth centuries, were the first of scientific consequence, and they form a useful starting point for briefly tracing the development of current views on the subject of primate intelligence.*

Thorndike's (1911) experiments were so devised that a hungry cat or dog had to work its way out of a cage closed by a catch operated from within, the stimulus to release activity being both the confinement itself and the sight of food which was placed near by. The experiment was performed in a reverse way when monkeys were tested. The animals had to open and enter the cage in order to win the food. As a result of a lengthy investigation, Thorndike concluded that cats and dogs learn by trial and error, that is, by the gradual elimination of random movements, and to explain this he defined his two well-known laws of effect and exercise—which, briefly summarized, imply that an animal will repeat (*a*) acts which give it satisfaction and (*b*) acts it has done before. While he believed that these laws also explain the learning of the monkey, Thorndike pointed out quite clearly that the monkey learns in a more efficient way than does the cat or dog. It can form more habits, it forms them more quickly, and the habits last longer. And in their quicker formation the monkey's habits often fail to bear the stamp of chance which characterizes trial and error learning, for after spending some time in useless random movement a monkey may suddenly go through a necessary series of acts as if they were part of an already established and adequate habit.

* For fuller discussion see Zuckerman, 1932*a*, Chapter X. The subject is also reviewed by Russell (1932).

Thorndike's psychological theory of animal learning received very strong support from the atomistic conception of behaviour that grew up on the physiological basis of the conditioned reflex. His views, however, no longer hold the field.

Their demolition did not come by way of a wider analysis of animal behaviour based upon more intricate experiments. It came as a more or less philosophical attack on the laws of learning which Thorndike had postulated, and as an attempt to show that cats and dogs were hardly the machines "trial and error" learning implied they were; that, on the other hand, they learn by the rational appreciation of the results of their actions, and by the application of insight. Unfortunately this criticism, which was mostly scholastic in character, had little effect until the publication by Adams (1929) of the results of a series of experiments identical with those made by Thorndike thirty years before. Adams' criticisms of Thorndike's experiments appear to be justified—and they are completely destructive. Thorndike's technique is described as deficient and his interpretation as faulty. Adams himself maintains that while cats may learn by trial and error—which occurs more commonly when the animals are excited and inattentive—they also learn through "insight", that is, by "looking over the situation" and, having by chance discovered the movable elements in the problem they are facing, smoothly going through the actions that bring success.

This is not the only way in which Thorndike's views have suffered. Criticism of recent years has amply demonstrated that the physiological prop to his laws of learning, the conditioned reflex, is weak, and that it cannot, however much it be assumed to be combined in "chains",* adequately explain intelligent behaviour. As Lashley (1929) has pointed out, the reflex theory's chief deficiency is its "implication of a point-for-point correspondence in the relations of receptors, nerve cells, and effectors." It is unnecessary to discuss here the destructive effect on such an hypothesis of the results of Lashley's own experiments.

Until the publication of Adams' paper, and in spite of all criticism, the view that the behaviour of monkeys and apes differed from that of other mammals by their manifestation of a characteristic called "insight" had steadily gained ground. Insight is not something that can be defined, but it is implied by the "appearance of a complete solution with reference to the whole lay-out of the field" of an experiment (Köhler, 1927). As the important distinction between primate and other mammalian behaviour, the concept had some validity, even though it defied scientific definition. Now that it has spread to the non-primate mammals, even to cats and rats, its scientific value seems questionable.

The popularization of this concept is the work of the Gestalt psychologists. Their main method of attack on the problem of animal learning is to devise simple experiments (rarely, if ever, involving surgical operations on the animals being investigated) in which, to give an example, food will be placed out of the reach of an animal, the only way to obtain the prize being a roundabout one—in the case of an ape it must be raked in with a stick, in the case of a cat a string tied to the food must be pulled. When the animal succeeds

* e.g. Pavlov's (1927) belief that in man "the different kinds of habits based on training, education and discipline of any sort are nothing but a long chain of conditioned reflexes."

quickly, it is said to display insight; if success is not so rapid as to merit this term, it is said to learn by trial and error. Obviously the term insight is no explanation of behaviour. It is simply an aid to description, which leads nowhere if too much reliance is placed upon it. To graduate insight, as Adams has suggested doing, into small and big insights cannot possibly help the scientific understanding of animal behaviour and learning.

This is not the place to enter into a discussion of Gestalt psychology, which is the school of thought to which many students of sub-human primate behaviour have belonged. It matters little here that some workers belonging to this school believe that the reactions of organisms are not composed "of isolated movements or a combination of discrete habits, instincts and wishes" (see Wheeler, 1929), or that they believe (see Russell, 1932) that animal behaviour can be explained scientifically simply in terms of conation and changing patterns or organizations of perceptive fields. What is important is that the results of such experiments on primate behaviour as have been carried out up to the present offer little or no support to the *a priori* belief that apes are more intelligent than monkeys. Thus a male capuchin belonging to De Haan (1931*a* and *b*) when tested by practically all the methods involving instruments which Köhler and others have used with apes, showed itself, in the view of its investigator, no less intelligent than the chimpanzee, and more so than the gorilla, orang and gibbon. Such a conclusion is obviously inconsistent with the view expressed by Yerkes and Yerkes (1929), who have suggested that the use of tools is a true indication of mental status, and who state that "manifestly instrumentation becomes increasingly possible and important between lemur and man. Probably it is one of the best indicators of mental status. Assuredly it places the anthropoid apes next to man in ability to achieve adaptation through modification of environment, and at the same time it indicates a great gulf between monkey and ape." Judging from the behaviour of professional performing animals, there is every reason to suppose that some catarrhine monkeys, if tested in the same experiments, would yield results as good as those of De Haan's capuchin.*

Capuchins fail to maintain their high level of per-

* Klüver (1933) gives a detailed account of a lengthy series of experiments designed to test the ability of a capuchin in the use of instruments, and his conclusions are in striking harmony with those of De Haan. He, too, expresses the doubt whether any wide gulf in "instrumental ability" separates the apes from monkeys, and he points out that in their skill with tools monkeys may differ as much among themselves as they do from apes. He himself discovered the common macaque to be far inferior to the capuchin in the use of instruments. On the other hand, he found that there was no such difference between these two species in experiments in which the animals had to learn to pull in one of two boxes characterized in a particular way. In some of these tests it became quite plain that the learning process is essentially the same in both types of monkey.

It would seem that any simple comparison of the "intelligence" of different sub-human Primates is more or less impossible, since a monkey's performance in one kind of experiment does not necessarily provide the basis for safely forecasting its reactions in another. This is clearly suggested not only by Klüver's experiments, and by the unequal performances of capuchins in tests with instruments and in tests of delayed reaction, but also by such experiments as those of Nellmann and Trendelenburg (1926), and of Guillaume and Meyerson (1930 and 1931) (to cite but a few authors); in some of these tests apes are much more successful than monkeys, in others much less so. With every succeeding investigation it becomes more difficult to see the criteria by which the relative "intelligence" of different sub-human Primates is to be judged. Indeed, it may prove impossible to devise as a test for this quality, a set of experiments that will both expose and neutralize the number of affective and other factors which, though probably external to the basic mechanisms involved in adaptive responses, nevertheless obviously influence the performances of monkeys and apes in experimental situations.

formance in "delayed reaction" tests. Thus they ranked immediately after lemurs in the fairly extensive series of Primates tested by Harlow, Uehling and Maslow (1932) and by Maslow and Harlow (1932) in respect to the capacity to respond correctly, after variable periods of delay, to one of two cups under which food had been placed. Baboons and mandrills were the most successful in these experiments—more so than were even some apes. An adult gibbon "gave responses definitely inferior to all of the adult baboons, and his performances were only superior to those of two of the eight macaques."

It is difficult to base any far-reaching generalizations on the results of these experiments, although Harlow and his co-workers, after admitting the existence of pronounced intra-specific differences in performance, due perhaps to such factors as age, motivation, and emotional characteristics, advance conclusions about the "superiority" of the baboon over other catarrhine monkeys, and of the gibbon's lack of superiority over the macaque. In view of the uncontrollable factors influencing individual success, as well as for other reasons, their attempt to project their findings into the realm of taxonomy and phylogeny cannot be regarded as a particularly happy one. The performances of monkeys in "delayed reaction" tests vary considerably according to the lay-out of the experiments,* and a great deal has still to be done before any final conclusions are drawn regarding the relative memory capacities of different Primates.

Few comparative studies of behaviour have included within their scope members of the Lemuroidea. De

* See Tinklepaugh (1932*b*) for a recent review of the subject. In his multiple delayed reaction experiments two chimpanzees gave better performances than two macaques.

[*Photo. by F. W. Bond*

21. CAPPED LANGUR AND YOUNG. Genus *Presbytis*

[*Photo. by F. W. Bond*

22. COLOBUS MONKEY. Genus *Colobus*

[*face page* 126

Haan (1930) has published the record of some simple experiments, which in their planning certainly leave much to be desired, and the conclusions to which they point, if indeed they point to any at all, are without doubt questionable. Food was hidden in full view of nine animals, representing the genera *Cebus*, *Macaca*, *Cercocebus*, *Cercopithecus*, and *Lemur*, in places such as the experimenter's pocket, or under a flower-pot. The capuchin, one of the macaques and the three lemurs always sought for the food, the others on many occasions losing interest the moment it was hidden from view. In the manner of their search, and in the success that attended their efforts, the lemurs showed themselves to be as skilful as the cercopitheque and one of the mangabeys. Klüver (1933), too, has investigated the behaviour of a lemur (*L. catta*), using a modification of the experimental lay-out which he devised for the determination of ranges of equivalent stimuli in monkeys. He found this animal far less efficient than monkeys, not only in the results, but also in the manner of its performances. The experiments of Fischel (1930) and Allesch (1931) also tend to show that in their adaptive responses lemurs are far inferior to monkeys.

* * *

Strangely enough, the present situation regarding "psychological" tests of sub-human Primates can be admirably summed up in the anti-Darwinian views expressed by St. George Mivart in 1873 (*b*). After referring to the differences in cerebral structure amongst Primates, this famous opponent of Darwin wrote:

> "Yet the psychical powers of different Apes are very similar. Not only the lowest Baboons of Africa (as e.g. the famed 'Happy Jerry' of Exeter Change) can be taught various and complex tricks and performances, but the less man-like American monkeys—the common Sapajous—are habitually selected by peripatetic Italians for the exhibition of the most clever and prolonged performances.
>
> "As to the two species of Sapajou, the brains of which are so different the one from the other, Professor Rolleston asks: 'Will anybody pretend that any difference can be detected in the psychical phenomena, the mental manifestations of these creatures, at all in correspondence or concomitant variation with their differences of cerebral conformation?'
>
> "The difference between the brain of the Orang and that of Man, as far as yet ascertained, is a difference of absolute mass. It is a mere difference of degree and not of kind.
>
> "Yet the difference between the mind of Man and the psychical faculties of the Orang is a difference of kind and not one of mere degree.
>
> "Thus, on the one hand we see that we may have great differences in brain development unaccompanied by any corresponding psychical diversities, and on the other we may have vast psychical differences which it seems we must refer to other than cerebral causes."

In spite of the excellent confirmation of Mivart's views provided by the recent psychological investigations of sub-human Primates, the belief that an ape is more intelligent than a monkey cannot necessarily yet be regarded as incorrect. It is possible that the usual methods of experimental psychologists are incapable of demonstrating the significance in overt behaviour of differences which are generally believed to exist in the complexity of the sensori-motor equipment of different sub-human Primates. Other kinds of investigation might succeed where these methods have failed, and, in particular, what is clearly needed at the present time is an adequately controlled comparative study of the learning capacities of the different animals in the order in a series of carefully selected situations. Before, however, discussing recent extensions of the experimental attack on primate behaviour, it is perhaps of interest to consider briefly those points relating to cerebral morphology which Mivart raised. In what significant ways do the brains of different sub-human Primates, particularly those of the sub-order Pithecoidea, differ from one another?

CHAPTER XI

THE BRAIN AS A MEASURE OF INTELLIGENCE IN THE SUB-ORDER PITHECOIDEA

IT is not proposed in this chapter to do more than touch on certain morphological points which seem to bear on the strange conclusion, suggested by experiments on behaviour, that monkeys and apes may be more or less equal in "intelligence." For a comprehensive account of the cerebral anatomy of Primates, reference must be made to such works as those of Elliot Smith (1902), Anthony (1916, 1917), Tilney (1928) and Le Gros Clark (1931, 1932*a* and *b*).*

The brains of the Primates—including, as Elliot Smith has shown, those of the Lemuroidea—are constructed on the same basal plan. Even the most primitive brain of this order that has yet been studied, the brain of *Microcebus murinus* (see Le Gros Clark, 1931), has a cortex which is considerably more differentiated than is that of brains of equal size belonging to mammals of other orders. It is from a brain like that of *M. murinus* that those of other Primates have probably evolved,† the fundamental changes being due, in the words of Elliot Smith (in Smith Woodward and others, 1920), to "the simultaneous cultivation of the

* Full references to the literature are given in these works.

† Le Gros Clark (1931) has speculated very convincingly on the more precise structure of the ancestral primate brain.

visual, auditory, tactile and motor areas of the cerebral cortex."

In its morphological effects this "simultaneous cultivation" shows itself as cortical differentiation and expansion. Actually the cerebral fissures and sulci characteristic of the order are believed to have developed as a mechanical result of this process of expansion, which seems to have been uniform throughout the group, since it is possible to state authoritatively that "an essential correspondence of fissural pattern exists in all primates" (Tilney, 1928). Differences in the sizes and proportions of any two primate brains may thus be regarded as representing differences between two definite stages of a uniform process, and presumably a brain intermediate in size between two such brains could be regarded as a stage of cerebral evolution intermediate between the stages they themselves represent. In a specific instance, the differences between the brain of man and that of the apes, in which all the human cortical formations are represented, may be regarded as bridged by the endocranial casts of such fossil remains as *Pithecanthropus* and *Sinanthropus*. The pre-frontal, parietal and temporal areas of these "Dawn Men" were far better developed than those of the apes, even though they were relatively much smaller than those of man of to-day.

Anatomists have been at pains to emphasize the proportional similarities between the brains of different Primates. For example, Le Gros Clark (1931), in drawing a picture of the orthogenetic trend of cerebral evolution in this order, writes of "the tendency among the various groups of the Primates to develop brains of similar proportions, with similar disposition of cerebral sulci, similar arrangements of nuclear masses, etc."

Recently Leboucq (1929) has suggested some interesting methods of giving a quantitative expression to these similarities, and to the relations of the parts of one primate brain to those of another. His cerebral coefficients are found empirically, and depend mainly on the determination of the surface area (including that hidden in the sulci), of both the whole cerebral hemisphere and of its separate lobes. His paper contains a full description of the chemical absorption and planimetric methods he followed in determining the extent of these areas.

Leboucq's coefficient K gives a measure of the relationship between the degree of convolution of the cerebral hemisphere and the degree of convolution of its lobes, the degree of convolution in each case being calculated as the ratio between the weight and the surface area. He gives the formula:

$$\frac{P}{p} = \frac{S}{s} \times K \text{ or } K = \frac{P}{p} \times \frac{s}{S}$$

where P is the weight of the hemisphere and p the weight of any one of its lobes, S the surface area of the whole hemisphere and s the surface area of the same lobe.

When comparing the lobes of a single brain the value of K will, therefore, be greater, the more extensive (convoluted) is the surface of a lobe relative to its weight. Thus K provides a good indication of the relative degree of surface development in any given part of the hemispheres. The following table of K values, which Leboucq provides, therefore suggests a constancy in the basic plan of the gross construction of the catarrhine brain:

TABLE VI

VALUES OF LEBOUCQ'S COEFFICIENT K

	Frontal Lobe. (Mean.)	Parietal Lobe. (Mean.)	Temporal Lobe. (Mean.)	Occipital Lobe. (Mean.)
30 men	0·95	1·05	1·12	1·46
4 chimpanzees . .	0·94	1·125	1·15	1·55
4 macaques . . .	1·025	1·085	1·06	1·55
3 baboons . . .	1·02	1·07	1·10	1·55

Leboucq gives another table to show the percentage contributed by each lobe to the total weight and surface of the cerebral hemisphere (the relation of these surface and weight figures gives the coefficient K):

TABLE VII

SURFACE AND WEIGHT PROPORTIONS OF PRIMATE BRAINS

	Frontal.		Parietal.		Temporal.		Occipital.		Insula.	
	Wgt.	Sfce.	Wgt.	Sfce.	Wgt.	Sfce.	Wgt.	Sfce.	Wgt.	Sfce.
30 men	37·8	36·0	21·8	23·0	22·1	25·0	9·5	14·0	8·8	2·0
4 chimpanzees . .	39·0	37·0	20·0	22·0	20·0	22·5	11·0	17·0	10·0	1·5
4 macaques . . .	30·0	30·4	18·5	20·8	23·5	25·0	14·0	22·0	14·0	1·8
3 baboons . . .	28·0	29·0	19·0	20·4	23·0	25·5	15·5	23·7	14·5	1·4

This table points to the interesting conclusion that the frontal lobe in the chimpanzee may contribute more to the total surface of the hemisphere than it does in man.

Leboucq has attempted to demonstrate another interesting coefficient of cerebral development. If the volume (V) of the brain is known, the radius of a perfect sphere of the same volume can be calculated, as can also the surface (S_1) of such a sphere. If S be the actual surface of the brain whose volume is V, then the coefficient R is $\frac{S}{S_1}$. The mean value of R in man is usually in the neighbourhood of 2·75. In the chimpanzee it is 2·28, in the macaque 2·26, and in the baboon 2·14. Leboucq points out that although the difference between the cortical development of the ape and of the monkey is not very great, the coefficients for the different Primates roughly correspond to prevalent ideas on the relative intelligence of the animals; this view, as the facts detailed in the last chapter show, probably needs qualification. Regardless of any possible correlation between this coefficient and intelligence, and allowing for experimental errors, the values of R, like the other figures which he has provided, once again seem to indicate a constancy in the construction of the brain of the Old World Primates.

Owing to the small number of specimens used in Leboucq's study, too strong emphasis cannot be laid on the actual data he provides. Apart from suggesting a useful quantitative method of studying the brains of different Primates, his figures do, however, definitely indicate that in gross form one catarrhine Primate's brain is very much like another, the inter-relations of surface to weight of the whole cerebral hemisphere and of its parts remaining more or less the same throughout this division of the order.* It is conceivable that the data Leboucq has adduced may have some relation to the conclusion, referred to in the preceding chapter, to which simple experiments on behaviour at present point.

So far I have considered only the external features

* Measurements of the surface area of some primate brains, and particularly of such parts as the area striata and the lateral geniculate body, are provided by Popoff and Popoff (1929). Their papers contain a résumé of previous work on this subject.

and the gross architecture of the brain. But very much the same picture of cerebral similarity, to which this external evidence points, is suggested by detailed anatomical study. Thus Tilney writes that in the internal structure of the brains of the Primates, "the homologous correspondence of each particular is pronounced beyond question of doubt", and he also remarks that "real difficulties might be experienced in distinguishing between the identifying characters of the human brain and the homologous structures in the brains of gorilla, of chimpanzee or of orang." Poliak (1932), who has made a very close study of the afferent fibre systems of the cerebral cortex of Primates, also writes that the brain of the lower Primates is "in its essential features and in its finer structure a simplified replica of the human brain." Of course these generalizations mask many neurological differences which do actually occur. For example, the direct pyramidal tract is ill-defined in macaques, but well-developed in chimpanzees (Leyton and Sherrington, 1917). Again, very interesting variations, as Le Gros Clark (1932*b*) has shown, occur in the lamination of the lateral geniculate bodies of different Primates. Moreover, the extent of the non-projection areas of the cortex varies in different Primates, a fact which Leboucq's work does not take into account. What precise functional significance all these differences may have is, however, still altogether obscure.

Cortical localization experiments have also revealed differences between different Old World monkeys and apes in the dynamic organization of the brain. Thus according to Leyton and Sherrington (1917) "discrete 'representation' of small local items of movement, each highly co-ordinated with others yet separably elicitable

. . . is more evident in cat and dog than in rabbit, more evident in the macaque than in cat or dog, in baboon than in macaque, in gibbon than in baboon, and in the chimpanzee, orang and gorilla than in gibbon."* Again one is left without any real knowledge of the bearing such differences have on the activities of these animals.

Since inter-specific and inter-generic differences in brain weight may conceivably have some bearing on the comparative study of the intelligence and learning capacities of different sub-human Primates, I have presented the data, without comment, in table VIII. Brain weights are given in absolute values, and also as ratios of body-weights, although, as many workers have pointed out,† this method of expressing the weight of the brain has few advantages. About half of the data considered in the table are records which I collected in the Prosectorium of the Zoological Society of London. The rest are taken mainly from Keith (1895) and from Hrdlička (1905 and 1925). A few are also quoted from Elliot Smith (1908) (12 records), and from the protocols of the Physiological Department of Yale University ‡ (6 records). The weight of the brain of *Tarsius* is taken from Le Gros Clark (1931). Except for two records of Gelada baboons, all the brain weights are from fresh, unfixed, material, and almost all are from animals believed to have been adult—although the very great variation in body-weight, and in brain-weight—body-weight ratios, suggests that this belief may not have always been well-founded. In a few cases I have in-

* See also Beevor and Horsley, 1890; Jolly and Simpson, 1907; and Brown and Sherrington, 1911.

† For recent discussion, see Harris, 1929.

‡ I am indebted to Dr. J. Fulton for permission to use these six records.

TABL

BRAIN-WEIGHTS AND BODY-WEIGH

Family.	Genus.	Species.	MALES: No. of Records.	MALES: Mean Brain-Weight. Gms.	MALES: Mean Body-Weight. Kgms.	MALES: Mean Brain-Weight—Body-Weight Ratio.	MALES: Brain-Weight Range. Gms.	MALES: Bod Weig Rang Kgm
Pongidæ	*Gorilla*	*gorilla*	1	426·25	90·72	4·7	—	—
	Pan	*satyrus*	—	—	—	—	—	—
	Pongo	*pygmaeus*	1	400·00	73·50	5·4	—	—
Hylobatidæ	*Hylobates*	*agilis* *lar* *pileatus* *leuciscus*	4	101·91	4·922	20·5	96·9–108·7	4·309–5
	Symphalangus	*syndactylus*	1	130·00	9·50	13·7	—	—
Cercopithecidæ	*Presbytis*	*entellus* *obscurus*	4	74·375	6·35	12·0	62·8–94·0	5·783–7
	Cercopithecus	*albogularis* *mona* *nictitans* *cephus* *sabaeus* *cynosurus*	8	66·70	3·70	19·7	55·0–80·0	1·915–5
	Erythrocebus	*patas*	1	113·00	4·412	25·6	—	—
	Cercocebus	*fuliginosus*	2	110·00	3·742	29·4	108·0–112·0	3·629–3
	Macaca	*radiata* *irus* *mulatta* *nemestrina*	13	86·77	5·38	19·0	69·7–103·1	2·041–7
	Cynopithecus	*niger*	1	97·50	3·45	28·3	—	—
	Theropithecus	*gelada*	1	168 (formalin)	10·004	16·8	—	—
	Papio	*cynocephalus* *papio* *anubis* *hamadryas*	15	175·40	12·970	19·0	142·0–213·0	4·080–2
Cebidæ	*Pithecia*	*monachus*	1	31·00	0·435*	71·3	—	—
	Saimiri	*sciurea*	—	—	—	—	—	—
	Cebus	*apella* *capucina* *fatuellis*	8	69·25	1·022	69·7	61·0–75·0	0·691–1
	Lagothrix	*humboldtii*	2	85·00	3·080	30·0	75·0–95·0	2·360–3
	Ateles	*geoffroyi*	4	90·875	2·044	44·9	82·0–95·5	1·835–2
	Alouatta	*seniculus* *inconsonans*	9	56·17	6·35	9·8	47·0–65·53	2·93–7
Hapalidæ	*Hapale*	*jacchus*	3	6·65	0·136	52·5	6·0–7·0	0·095–0
	Oedipomidas	*sp.*	1	10·20	0·2655	38·5	—	—
	Leontocebus	*rosalia*	1	13·00	0·335*	38·8	—	—
Tarsiidæ	*Tarsius*	*spectrum*	1	3·6	0·092	39	—	—
Lemuridæ	*Lemur*	*variegatus* *macaco* *fulvus* *mongoz* *catta*	8	29·70	2·335	13·5	21·8–39·0	1·390–3
	Microcebus	*coquereli*	—	—	—	—	—	—
Lorisidæ	*Nycticebus*	*coucang*	—	—	—	—	—	—
	Perodicticus	*potto*	1	15·60	—	—	—	—
Galagidæ	*Galago*	*garnetti*	—	—	—	—	—	—
Daubentoniidæ	*Daubentonia*	*madagascariensis*	—	—	—	—	—	—

* Maturity uncertain.

III

ADULT SUB-HUMAN PRIMATES

…rain-eight—Body-Veight Ratio Range.	Source of Data. (see text)	FEMALES No. of Records.	Mean Brain-Weight. Gms.	Mean Body-Weight. Kgms.	Mean Brain-Weight—Body-Weight Ratio.	Brain-Weight Range. Gms.	Body-Weight Range. Kgms.	Brain-Weight—Body-Weight Ratio Range.	Source of Data. (see text)
—	K.	—	—	—	—	—	—	—	—
—	—	2	373·50	31·75	11·9	368·0–379·0	—	—	K.Z.
—	K.	1	314·00	—	—	—	—	—	Z.
7–22·3	H.K.	5	92·30	5·78	16·3	78·0–96·4	4·76–7·25	13·3–20·2	K.
—	K.	—	—	—	—	—	—	—	—
2–16·2	K.Z.	7	62·90	5·046	12·7	56·7–72·0	4·53–5·90	10·9–16·9	K.
4–30·0	Z.H.	9	61·00	3·391	19·4	51·0–82·0	1·757–5·082	11·8–34·1	Z.
—	Z.	1	86·00	3·912	22·0	—	—	—	Z.
0–30·9	H.	3	101·93	5·052	23·0	98·78–105·0	3·175–7·560	13–33	H.Z.
4–49·0	H.K.Z.F.	13	76·00	3·55	24·3	59·0–110·0	1·165–6·151	14·5–42·9	Z.K.H.F
—	H.	1	84·00	3·18	26·4	—	—	—	Z.
—	F.	1	187 (formalin)	9·54	19·6	—	—	—	F.
4–43·4	H.Z.F.	6	157·00	7·725	21·7	130·0–167·0	5·443–11·226	14·8–29·8	H.Z.
—	Z.	1	36·22	0·538	67·3	—	—	—	K.
—	—	3	19·00	0·311*	62·8	18·0–20·0	0·300–0·322	59·0–66·6	Z.
5–88·3	Z.H.	4	63·50	1·528	42·2	52·0–72·0	1·247–1·814	35·8–52·1	H.Z.
7–40·3	Z.F.	—	—	—	—	—	—	—	—
2–49·7	H.	1	94·50	2·268	41·7	—	—	—	H.
6–16·0	H.	2	46·25	1·947	28·8	44·0–48·5	1·170–2·725	16·1–41·4	H.
3–73·2	H.Z.	4	7·00	0·223	33·4	5·0–8·0	0·152–0·265	21·3–52·6	Z.
—	H.	2	8·925	0·325	27·4	8·0–9·85	0·290–0·360	27·4–27·6	H.
—	Z.	1	12·00	0·289*	41·5	—	—	—	Z.
—	C.	—	—	—	—	—	—	—	—
7–20·8	H.S.Z.	13	25·484	1·813	15·0	18·0–32·0	0·940–2·720	10·2–20·7	H.S.Z.
—	—	1	8·00	0·389*	20·6	—	—	—	Z.
—	—	3	10·33	0·452*	23·0	9·0–11·0	0·405–0·480*	19·0–27·2	Z.
—	Z.	—	—	—	—	—	—	—	—
—	—	2	9·50	0·590*	16·3	9·0–10·0	0·540–0·640*	14·0–18·5	Z.
—	—	1	36·75	1·235	29·8	—	—	—	Z.

cluded records, from my own series of data, of animals that were not definitely known to be adult (as judged by dentition and signs of reproductive maturity), and these have been indicated by an asterisk. Brain-weight—body-weight ratios are given in the form

$$\frac{\text{Brain-weight}}{\text{Body-weight}} \times \frac{1000}{1}$$

This ratio for adult human beings is usually in the neighbourhood of 28 (Quain).

The total data are very few, and particularly few in the case of the apes. For these Keith (1895) gives a table of weights taken mostly from preserved specimens. The average brain-weight of twelve adult male gorillas was 463 gms., of six females, 450 gms.; of two male chimpanzees, 406 gms., of four females, 393 gms.; of three male orangs, 431 gms., and of one female, 393 gms. The average brain-weight—body-weight ratios for the males were 5·1, 5·1, and 5·4 respectively.

It is almost unnecessary to state that I have made only a partial survey of the cerebral characters of the Primates. The conclusion to which the data I have considered point is that known morphological differences in the essential structure of the brains of the catarrhine group have not as yet been shown to have far-reaching significance in the problem of learning. These data were not in any way selected. On the contrary, I have searched for facts which would demonstrate differences that could without question be regarded as of real importance from the point of view of the evolution of intelligent behaviour. The fact that the data considered are not known to imply such a conclusion may either be taken to support the anomalous thesis with which this chapter was begun—namely, that monkeys and apes are more or less equal in "intelligence"—or, alternatively, that such methods of comparative psychology as have hitherto been applied to sub-human Primates must be more closely linked with those of cerebral anatomy if either method of research is to become capable of defining real differences in intelligence which may actually exist, and which are suggested by the normal behaviour of the different animals themselves.

CHAPTER XII

THE PHYLOGENETIC IMPLICATIONS OF CORTICAL PHYSIOLOGY

An outstanding and elaborate study of the cortical physiology of sub-human Primates has recently been published by Fulton and Keller (1932). Primarily the study concerns the Babinski reflex, which is a form of plantar response. When the sole of a healthy adult man or woman is stroked, the big toe automatically turns down. This is the normal plantar response. In young infants and in adults suffering from some lesion affecting the motor pathways from the brain to the foot, the toe responds to such stimulation by bending upwards. It is this abnormal response that is termed the Babinski reflex. Clinical and pathological work has established the conclusion that the manifestation of such a reflex as an isolated phenomenon in human infants and in human adults implies the non-existence and the interruption respectively of a particular kind of connection between the cortex and the foot.

Using monkeys and apes as experimental subjects, Fulton and Keller studied the effects which various experimental lesions to the spinal cord and brain have on this reflex. Their main results may be summarized as follows. For the purpose of this summary it should be clearly understood that the elicitation of the reflex in an ape or a monkey after the experimental interruption of its cortical-foot pathways implies the previous existence in the animal being investigated of the specific cortical-foot relation under consideration. If the reflex does not appear, it follows that the cortex and foot were not connected in the particular way implied by the Babinski reflex.

A Babinski response does not follow the unilateral removal of all pyramidal control in monkeys.* It is observed only when the lower lumbar segments are completely freed from the influence of higher parts of the nervous system, and then it occurs as part of a generalized flexion reflex, not as a simple isolated phenomenon.

Baboons † show themselves to be somewhat similar to monkeys in these experiments, but they also differ in some important points. For example, removal of a cerebral hemisphere,‡ or the localized extirpation of the cortical leg area,§ produces a primitive form of Babinski response, and after such lesions recovery of motor power is slower and less complete than in monkeys. This suggests that in these animals the cortex has assumed controlling functions of the foot which are lacking in the more arboreal members of the *Cercopithecidæ*.

A still more conspicuous neurological difference between monkeys and baboons is found in the baboon's exhibition of a response to which Fulton and Keller give the name "monoplegic flexion". This reaction

* *Macaca mulatta*, *Cercocebus torquatus*, *Cercopithecus sabæus*, *Cercopithecus diana* and *Erythrocebus patas*.

† Three species of *Papio* and two animals of the species *Theropithecus gelada*.

‡ A single experiment with a somewhat equivocal result (see Fulton and Keller's description, 1932, p. 55).

§ Seven experiments.

comprises the withdrawal of a paralysed leg (paralysed as a result of removal of its cortical motor area), on the application of a nocioceptive stimulus anywhere on the body, which suggests, as Fulton and Keller write, that "the reflexogenous zone had spread from the sole to include the entire body surface". Such a response was never more than "poorly developed" in monkeys that had been treated in the same way.

The results of some similar experiments with a gibbon (*Hylobates lar agilis*) showed (*a*) that the return of voluntary power after injury to the motor area of the cortex is much slower than in either the baboon or in other monkeys, and (*b*) that in this animal a Babinski response appears shortly after ablation of the cortical-foot area of the side concerned. This response disappeared after the return of voluntary power to the monoplegic leg, but when, seventy days later, the opposite cortical-foot area was removed, the Babinski sign reappeared "in accentuated form on the homolateral side as well as on the opposite side", a fact which indicates, as Fulton and Keller point out, "some degree of bilateral representation in the gibbon's cortex." After this second cerebral injury the Babinski sign became a permanent reflex.

The reactions of chimpanzees to these and similar experiments were even more different from those of monkeys than were those of the gibbon. A Babinski response appeared almost immediately after cortical ablation of a foot area, and about a fortnight after the operation the reflex had begun "to assume characteristics similar to the corresponding response in man." It persisted long after voluntary movement in the leg had returned. Removal of the opposite cortical foot area at a later stage produced the same effects in the opposite limb, and greatly intensified the Babinski response of the originally affected side. Moreover, a permanent change took place in the posture of the ipsilateral foot. In a certain experiment, the cortical foot area of one side was first removed, and forty-three days later the entire cortical area of the opposite lower extremity was ablated. The immediate result of the second experiment was that both extremities became flaccid, all reflexes at the same time being suppressed. "This indicates clearly", as Fulton and Keller remark, "that some degree of bilateral representation exists in the foot area of the chimpanzee."

In interpreting their results, Fulton and Keller indicate two wider problems on which their experiments bear—firstly, the problem of the evolution of cortical dominance, and secondly, the questions of primate taxonomy and phylogeny. The bearings of their investigations on the first of these problems are of far-reaching importance, and are referred to in the next chapter; those relating to the questions of primate taxonomy and phylogeny are somewhat debatable, and are considered here.

In an earlier chapter (pp. 20–23) I discussed the arrangement of the genera of the family *Cercopithecidæ*, pointing out that the development of the evolutionary doctrine had resulted in a false interpretation being made of the traditional order of genera which post-Darwinian systematists had inherited from their pre-Darwinian forerunners. Thus I noted that baboons are often referred to as being lower in the evolutionary scale than are macaques, actually without any real justification, and presumably because they come first in the generic list of *Cercopithecidæ*. Since this belief in a phylogenetic series of monkeys is widely prevalent, it is not surprising that Fulton and Keller, who approach the problem of the systematics of Primates from the physiological angle, write about the various catarrhine species as though they actually do represent some kind of evolutionary series. Thus, for example, these authors refer to *Erythrocebus patas* as being "more lowly" than *Cercocebus torquatus* (= *lunulatus*) and to this species and *Macaca mulatta* as being "higher monkeys" than the cercopitheques. And they write, "The gibbon and siamang are generally regarded as the most lowly of the anthropoid apes, but since there is disagreement as to definition of both 'anthropoid' and of 'ape', it is perhaps better to state that the gibbon and siamang are of the lowest order of tailless primate",*—in this case "low" and the implied "high" presumably having man as a point of reference.

In itself such description cannot be objected to, but since Fulton and Keller also attempt to rearrange the species of the *Cercopithecidæ* into a phyletic order more consonant with their experimental results, it is necessary to examine the bearings of their conclusions on the systematic distribution of other taxonomic characters.

Their consideration of the baboon demands close examination. They write that if an "order" intermediate between the monkeys and apes is to be recognized, it should contain the baboons alone. One of their arguments in favour of this contention is the fact that the baboon has a brain-weight very nearly twice that of any other member of the family *Cercopithecidæ*. A difference of this kind, however, is not usually

* "Tailless primate" is not a more convenient term than anthropoid ape, since the terms would include such forms as the Barbary ape, *Macaca sylvana*, and the Celebes ape, *Cynopithecus niger*.

regarded as important in mammalian systematics. Moreover, if brain-weight is expressed as a fraction of body-weight, there is no significant difference, as table VIII shows, between the baboon and some other members of the family in which it is at present classified. Fulton and Keller also place great emphasis on the physiological findings which support their view. Thus the baboon's motor cortex is more highly differentiated than is that of the macaque. For example, it possesses a much greater cortical hand representation, and when stimulated in a certain area, raising and turning the tip of the snout can be elicited—a reaction seen in modified form in the gorilla but not in the "lower monkeys". To these considerations Fulton and Keller add facts about the baboon which they themselves have discovered, namely, the existence of a certain amount of bilateral cortical representation in its brain, and the animal's slow recovery of voluntary power after cortical injuries.

But the observations bearing on these points are few, and, as Fulton and Keller are careful to indicate, "variations in any one species, due to age, completeness of lesion, etc., are considerable." For this reason, it is very questionable whether the available facts of cerebral physiology provide enough justification for separating the baboons into a separate family. Over and above this consideration, there are very sound morphological and physiological reasons, into which it is unnecessary to enter here, for continuing to classify the baboons as a genus of the family *Cercopithecidæ*.

Before discussing any further the question, raised by Fulton and Keller, whether it is possible to indicate evolutionary considerations in the classification of the Old World monkeys, it is interesting to consider a suggestive conclusion to which their experiments point, but which they themselves do not mention. This conclusion is that " cortical dominance " (as measured by the Babinski reflex, bilateral cortical representation and slower recovery rates) may be closely correlated with a terrestrial habit, or, more correctly, that it is a phenomenon unrelated to the normal mode of progression of true arboreal monkeys. Thus it is evidenced by the baboon, which is fundamentally a terrestrial animal. It is also shown by the Gelada—*Theropithecus*—to which Fulton and Keller refer as a baboon, but which Garrod's (1879) and Pocock's (1925*c*) researches plainly show is not a baboon. This animal is also terrestrial. The gibbon and the chimpanzee again, which demonstrate a greater degree of such dominance than do either true baboons or the Gelada, are regarded by some students (e.g. Pocock, 1929) as fundamentally terrestrial animals, which have only secondarily returned to the trees. Those who would not subscribe to so extreme a point of view are nevertheless usually careful to point out, as Gregory (1922) has done, that the locomotion of the apes, including the gibbons, is not "quadrupedal in the primitive way", and that the methods of progression of these animals are totally different from the locomotor habits of truly arboreal monkeys. It is thus conceivable that there is some relation between modes of locomotion and cortical dominance, perhaps even some relation between life on solid earth and the inherent tendency of the primate stock to relegate to the highest parts of the brain functions which in truly arboreal forms are subserved by lower centres.

At this point the taxonomic question raised by Fulton and Keller may again be considered. Is it possible to indicate phylogenetic considerations in the classification of such a family as the *Cercopithecidæ*, where different genera have evolved on paths which, from the point of view of form and habitat, are so far apart? Is it, for example, possible to indicate the nature of the phylogenetic relationship of such different types as the baboon, an animal weighing about seventy pounds, which in the greater part of its distribution in Africa lives on arid rocky hills, and an animal like the West African Talapoin tree-monkey, which weighs no more than two and a half pounds?

To some extent it is possible. In spite of the great diversity of the animals comprising the *Cercopithecidæ*, they can be arranged, in accordance with the distribution of structural and physiological characters, in smaller groups in which genera are probably more closely related to each other phylogenetically than they are to the members of other groups. Thus the drills, mandrills and baboons form one large division; related to these animals are the mangabeys, and to them probably the macaques. The cercopitheques and Patas monkeys form a separate group, which superficially seems related to the sub-family *Colobidæ*. These constitute the conspicuous sub-groups of the family, but they do not account for all the species which are justifiably included in the *Cercopithecidæ*. Thus there remain the Gelada baboon and the Talapoin monkey, both of which appear to be fairly isolated types in which are conglomerated strange mixtures of cercopithecoid characters. Unfortunately the phylogenetic significance of these combinations of characters is not clear. As the facts of palæontology, morphology and physiology show, existing members of the *Cercopithecidæ* are very probably the survivors of a number of types which differentiated from one basal stock, and the more conspicuous of the sub-groups of the family probably diverged very early from this main stock, each group then evolving separately. But this is as far as available facts will allow one to speculate. Data considered in earlier chapters of this book provided very suggestive evidence that the same characters have evolved separately in different groups of Old World Primates, and the possibility of manifold parallel evolution within the family is an obstacle to the tracing of any detailed phyletic scheme.

CHAPTER XIII

THE EVOLUTION OF PRIMATE BEHAVIOUR

(*a*) Physiological

CORTICAL dominance reaches its highest expression in man. The varying degrees to which it demonstrated itself in the monkeys and apes of Fulton and Keller's investigations may be regarded as stages in one particular cortical evolution. From this experimental point of view the degree of evolution may be measured by the rate of recovery of motor power after cerebral operations, by the extent to which such operations depress spinal reflexes, and by the amount and type of injury that is necessary to produce the Babinski sign. The evidence provided by Fulton and Keller shows that the cortex is of greater functional significance in some motor activities of the gibbon and chimpanzee than in corresponding activities of monkeys. Since a true Babinski sign does not appear in monkeys after destruction of the main cortical pathways (the pyramidal tracts), it may be assumed that in them the cortex has not taken over certain controlling functions of the foot as it has done in the ape. Again, as the following table demonstrates, the rate of recovery from cortical lesions is quicker in monkeys than in apes:

TABLE IX

"*Time required for recovery of reflexes and voluntary power in various primates following removal of the cortical representation of the lower extremity.*

(Since the plantar reflexes are the first to reappear, the first column indicates the duration of complete monoplegia; numerals indicate days.)"

	Plantar Reflexes.	Knee-jerk.	Return of Voluntary Power (Flexion).			
			Hip.	Knee.	Ankle.	Toe.
Macaque * . .	†	†	1	2	3	5
Baboon * . .	0·5	†	2	3	6	10
Gibbon . . .	1·3	3	4	6	21	35
Chimpanzee .	2	3	2	4	25	?40–60

* Average of four animals.
† Greatly depressed during first 24 hours.
Table from Fulton and Keller, 1932.

Further evidence of the evolution of cortical motor function can be found. Bilateral representation of the leg area, although minimal in most monkeys, is increasingly present in a series formed by the baboon, gibbon and chimpanzee. Again, the physical state produced by a cortical lesion in apes is much more like spinal shock than is the condition that a corresponding lesion produces in monkeys. In Fulton and Keller's own words, "as one ascends the primate scale the state of monoplegia observed after a lesion of the motor cortex comes to simulate more and more the condition of reflex depression seen after transection or semisection of the spinal cord." These experimental observations are all indicative of the process of encephalization, of the assumption by the cortex of new controlling activities.

It must of course be remembered that experiments

such as these constitute only the beginnings of the exploration of the cerebral cortex of sub-human Primates, and that in themselves they do not provide the material for far-reaching conclusions about the evolution of general cerebral function. It may still bc shown that the cortex subserves certain motor functions in cercopitheques, macaques, and mangabeys which it does not do in animals such as the chimpanzee, gibbon, and baboon, in which it controls the Babinski reflex.*

The process of encephalization is generally believed to go hand in hand with more highly differentiated cortical localization of function, the term being used to imply far more than just motor function. In rats, as Lashley (1929) has shown, such localization is of a very low order, and the specificity of the cortex is very slight. He has presented strong evidence, based upon a large series of experimental cortical ablations, that the rate at which these animals learn, apart from problems based on brightness discrimination, is directly proportional to the amount of cortex they have left after operation—the site of the cerebral lesion being relatively unimportant. Lack of cortical specificity is equally well shown when the animals are operated on after they have been previously trained to a task, for the habit is affected by any cortical lesion above a certain size. Once again the disturbance is proportional to the amount of destruction. As Lashley himself states, it is probable that—"all parts of the cortex participate equally in the performance of the habit and that lesions of equal size produce equal loss of the habit, irrespective of their locus." He also writes: "The

* For further reports on the functional investigations of the cortex in sub-human primates, see Fulton, Jacobsen, and Kennard (1932), Jacobsen (1933), Fulton and De Barenne (1933), and Bieber and Fulton (1933).

[Photo. by F. W. Bond

23. SIAMANG GIBBON. Family *Hylobatidae*

[Photo. by F. W. Bond

24. GORILLA. Genus *Gorilla*

[face page 150

increase in efficiency of learning with increasing amounts of cortical material . . . results from an increase in the amount or intensity of some qualitatively unitary thing which contributes to the efficiency of a variety of functions," and further that "for the efficient performance of the maze habit, it is not necessary that any two cortical fields shall be in direct connection with one another." His work has clearly shown that however specific and integrated cortical *functions* may be in the rat, functional differentiation is largely independent of any kind of *structural* differentiation.*

With his critical powers sharpened by these experimental observations, Lashley reviewed the evidence on which the belief in a general evolutionary tendency to an increasing specialization and localization is based. His conclusion is noteworthy. In his view, apart from the function of spatial orientation (i.e. the precise differentiation of the sensory and motor projection fields of the cortex), "there is little evidence of a finer cortical differentiation in man than in the rat." Intelligence he would explain not as the function of a particular cerebral centre, but as a dynamic function of the entire cortex. This opinion is directly opposed to the view that the human frontal lobe is a centre which is concerned with "the general orderly co-ordination of psychic processes" (Bolton, 1903), or an organ which, in Tilney's words (1928), is "the accumulator of experience, the director of behavior, and the instigator of progress."

Lashley's opinions receive strong support from certain clinical observations made during the war. They are also supported, particularly in their opposition to

* See Lashley (1933) for a recent statement of his views on the integrative functions of the cerebral cortex.

the view that the frontal lobes are the main seat of intelligence, by the results of Jacobsen's experiments on Rhesus monkeys (1931). In these experiments animals were first trained in certain sensori-motor habits, and various amounts of cerebral tissue were then removed. After recovery, the monkeys were re-tested to discover the degree of their retention of these habits, and their capacity to learn new ones. Ablations were carried out only in the areas anterior to the pre-central sulci (in the region homologous with man's frontal and pre-frontal areas) and in the parietal region; the primary object of these experiments was to determine "whether or not a particular part of the cerebral cortex (the frontal lobes) is more essential for the formation and retention of sensori-motor habits than are certain other parts of the cortex." The results show that injuries to the frontal and pre-frontal areas, while they do not affect the retention of simple problem-box habits or pattern-discrimination habits, do impair more complicated problems involving the integration of a number of simple latches. Lesions in the parietal areas produced similar results. On the basis of these experiments Jacobsen concluded that the frontal lobes could hardly be regarded as the organ of intelligence, pointing out that his observations support the view "that intelligent behavior is the product of the equilibrated action of the entire nervous mechanism."

As with experiments on the motor functions of the cortex, it should not be forgotten that experiments on the cortical basis of intelligence, such as those referred to above, are very much in their infancy. When Lashley writes that there is little evidence of cortical differentiation either in the rat or man, he is not considering anatomical localization with respect, for example, to the connections of specific parts of the cortex with different parts of the different thalamic nuclei (e.g. see Le Gros Clark, 1932*c*), and his experiments do not expose any functional significance which such localization may nevertheless have. Because of this, it is still perhaps necessary to regard with caution the wider implications of this field of experimental study.

(*b*) Cultural

If the various facts referred to in this and the preceding three chapters are taken at their face value, they lead to strange conclusions. The kind of psychological experiment that has been employed up to the present has succeeded in demonstrating that species of the sub-human Pithecoidea as far apart as *Cebus hypoleucus*, a South American capuchin, and *Pan satyrus*, the chimpanzee, succeed more or less equally well in experimental tests which are regarded as providing a criterion of mental status. As I have tried to indicate elsewhere (Zuckerman, 1932*a*), this equality covers far more than the activities of Primates in captive and experimental conditions. On the basis of our existing knowledge of the organization of the social lives of animals so unlike each other morphologically as gorillas and macaques, it is impossible to define differences which would be indisputably regarded as significant from the point of view of human social evolution. For example, even though the behaviour of apes may superficially seem to be more man-like than is that of members of the *Cercopithecidæ*, we do not yet possess evidence that the apes comprehend the significant characters of different social relations any better than do the baboons,* who in their social responses show that they do not distinguish clearly and permanently between the sexes, between young and old, between living and dead, between homosexuality and heterosexuality, between monogamy and polygyny.

As I have already suggested, however, naturalistic observations and experiments of a purely psychological kind such as those briefly referred to in Chapter X may be far too crude to expose the real differences that possibly exist in the intelligent responses and learning capacities of monkeys and apes, and which possibly represent phases in the evolution of intelligence. The usual preoccupation of these types of study may in fact not be of a kind that can provide knowledge about the real nature of such an evolution, or of the physiological nature of the differentiations in the behaviour of these animals.

Even where differences in the dynamic construction

* For further information on this subject, see Zuckerman, 1932*a*, Chapters XVIII and XIX. After this book was in the press, I learned of certain parallel observations on apes, on record in the Department of Psychobiology of Yale University. The third baby of a chimpanzee (Mona) died within twenty-four hours of its birth. The mother guarded the corpse zealously, and for a month resisted all efforts to remove it, carrying the decomposing remains wherever she moved. A short time before it was taken from her, she was seen to crack the skull open with her teeth, and to eat some of the contents. Once it was removed, she seemed oblivious to her loss. The same unconcern is shown by mother chimpanzees whose living babies are taken from them. Though the greatest opposition has to be met by the experimenter anxious to effect the separation, the mother animals seem completely adjusted to the loss of their young within a few moments, and in a very short time they give no sign of recognition when they are brought face to face with their abducted offspring. Similar behaviour is also characteristic of baboons and monkeys, and it would seem to typify a level of development in which the amount of differentiation of social stimuli and situations falls far below that met with even in the most primitive human community. I am indebted to Dr. O. L. Tinklepaugh for permission to quote the data about the chimpanzee Mona.

of the cortex of the brains of different sub-human members of the catarrhine division of the Primates are revealed, they appear, however, to have little value in the natural economy of monkeys and apes in general. Chimpanzee types and cercopithecoid types apparently evolved in much the same period of the earth's history, and both have survived in arboreal surroundings until the present day. Neither the one nor the other can be regarded as the more successful zoological form—for no standard is yet available by which to make a judgment of their relative efficiencies within the more or less common environments in which they live. Perhaps, however, a more highly differentiated cortex with respect to motor functions could find overt expression in more varied play activity, and it may actually be that the play activities of chimpanzees or other apes are more varied than those of monkeys, and that this level of their activity is representative of a level out of which developed the more organized and the more defined use of implements. Possibly, too, as I have already pointed out, a more highly differentiated cortex is simply a specialization manifested by types which are fundamentally terrestrial in habit. In either case, it is difficult to see of what specific account cortical differentiation in respect to motor functions will be in the problem of the evolution of human intelligence, if Lashley's conception of the cortical basis of intelligent behaviour proves to be correct.

Human beings had been living on earth for an unknown but, historically speaking, vast period before some few (probably less than seven) thousand years ago there developed the complex culture involving agriculture, animal domestication and metal-working, which persists as the foundation of the world's present material civilization. Before this, from the time of the emergence of *Homo*, men had been food-gatherers, and to paraphrase Elliot Smith (1930*a*), the majority were content with the roughest of shelters and clothing. Crafts were restricted to the making of a few rude implements, and social life was one of small groups. And yet, if endocranial casts are a guide, the people of the Upper Palæolithic who lived in this way had brains which were no different in structure from those of modern civilized man. The people who live like this to-day also have the same brains; in the space of a few generations, however, they are capable of undergoing a cultural transition from primitive nomadism to organized industry.

The behaviour of apes can reasonably be compared with that of man living in this primitive collecting phase, and several very significant differences can readily be recognized. There are man's cultural instruments—speech, fire, tools, etc.; there is his omnivorous diet; and, to choose one more point, there is man's monogamous habit. The last, biologically speaking, may sometimes be more apparent than real. Most food-gatherers that have been observed (see Elliot Smith, 1930*a*) live in monogamous family parties, and very likely in some, perhaps in most, cases this implies a lasting marital habit; in others, it need mean no more than that in these groups a man lives with only one woman at a time.

These attributes of primitive man are characters in which he differs from the apes, and judging by their distribution amongst still existing and recently surviving food-gatherers, they are essentially characters which allow one to distinguish a human from a sub-human level of development. It is probable that their origin depended largely upon the evolution of an essentially human neurological equipment. It can readily be argued that external circumstances also played a big part in their emergence.

Unless man's pre-human ancestors differed fundamentally from other catarrhine Primates their social lives were overtly organized according to a scale of dominance. It may therefore be assumed that they lived more usually in polygynous, and not in monogamous family parties (monogamy appears to exist amongst monkeys and apes as a rule only when personal force is not strong enough to allow of polygyny (Zuckerman, 1932*a*)). Another fairly safe assumption is that the diet of these pre-humans, like that of monkeys and apes, was mainly frugivorous.* (In captivity monkeys and apes will sometimes eat such foods as eggs and worms, and occasionally, but very rarely, meat. It is possible that a little animal food also forms part of the natural diet of these animals.)

Meat is a usual part of human diet, not only in civilized societies, but also in primitive food-gathering communities. The evidence we have of palæolithic man in Europe and China shows conclusively that he too was predominantly a hunter. It also suggests that he was usually a cave-dweller. This supposition is supported by the fact (see China and Uvarov, 1932) that many of the animals, such as bugs and crickets, now associated with man in his immediate environment, actually belong to the ecological group of rock and cave communities.

It has often been speculated that man's immediate forebears changed from a predominantly frugivorous to an omnivorous diet as a result of environmental

* See on this subject Nissen, 1931, and Bingham, 1932.

changes. The reduction of forests in a period of aridity,* and a consequently dwindling food supply of the kind to which anthropoid apes are adapted, may have acted as a selective factor in determining the evolution of a group of emergent men with more plastic food habits than those, for example, of existing apes. Whether this change in diet was related to an environmental "accident" of a kind such as the one just described, or whether the pre-humans who had diverged from an ape-like stock through a process of mutation and selection migrated, "for no reason at all", from the forest belt into the open plains, where perforce they had to develop carnivorous tendencies, it was in all likelihood correlated with the repression of such an overt manifestation of dominance as polygyny. Several ways in which this may have come about can be visualized. Conditions may have been too stringent to allow of family parties as large, for example, as those of gorillas living in rich arboreal surroundings. The transfer from a collecting to a hunting economy may itself have had far-reaching social and sexual consequences. Thus sexual division of labour in the collection of food is not known to occur amongst sub-human Primates. Indeed, the evidence is almost conclusive that female monkeys and apes, whether non-pregnant, pregnant, or nursing, forage for their own food. So, too, do the young animals. A social life organized on these lines could hardly have been maintained by the semi-carnivorous animals that emergent men probably were, since it is altogether unlikely that either females, whether non-pregnant, pregnant, or parturient, or their

* The opening phases of the Pliocene are believed to have constituted such a period in the "vast central Pal-Asiatic region", where many authorities believe man to have originated (see Black, 1925).

young would be capable of hunting for themselves. Hunting as an occupation of women is practically non-existent among the peoples of the world. It is through the activities of their males that the women of food-gathering communities are supplied with meat. Economic specialization of the sexes may therefore have begun in the retention of collecting by the females, and the development of hunting by the males, who were far better fitted physically for such an occupation.

On *a priori* grounds it can be argued that the expression of overt polygyny declined from the moment when primitive man became predominantly a hunting and food-sharing animal. On the one hand, it is simpler for males to provide for a small family than for a large one; on the other, polygynous animals could not have gone hunting if in their absence their females were habitually abducted by males not so fortunate in the possession of mates. A process of selection operating through such social channels may very well have predisposed to the development of an overt monogamy, and to the clear and permanent realization of the significance of social relations. Conceivably, too, such a process may have been linked with a more developed sexual selection than is apparent among sub-human Primates.

It thus appears likely that the transition from an ape-like to a human mode of social life, one of the most important phases in man's mental development, may have been largely a matter dependent in the first place upon circumstances arising from without, rather than upon circumstances depending on the actual structure of the brain. This possibility clearly suggests the importance of developing studies on the palægeography of extinct Primates. It also indicates the necessity for ecological research which would include not only studies of the feeding habits (and alimentary physiology) of still existing Primates, but also investigations of the various problems, referred to in Chapter IX, relating to migration, isolation and species cohesion.

CHAPTER XIV

FUNCTIONAL DIFFERENTIATION IN RELATION TO THE EVIDENCE OF MORPHOLOGY AND PALÆONTOLOGY

THE facts regarding the generative functions, the blood, purine metabolism, perceptor organs and behaviour patterns, which have been discussed in preceding pages of this book, are mostly consistent from the taxonomic point of view. It would therefore be interesting to bring these data together in this concluding chapter and, although this necessitates brief recapitulation, to examine their phylogenetic significance, especially in relation to the palæontological and morphological evidence bearing upon the evolution of the Primates. I shall not include in this "summing up" the data considered in the chapter on the parasites of Primates, as in the present state of our knowledge it is perhaps inadvisable to mix with the anatomical, physiological and behavioural characters of animals, characters which to some extent have an independent existence. Other data discussed earlier in this book, and to which I shall not refer, are those relating to hybridization, to species cohesion, and to stages in the evolution of primate behaviour. These questions have been omitted not only because the available facts regarding them hardly allow of consideration in a phyletic scheme, but also because to some extent they have already been summarized and brought together in preceding pages.

The following are the main points which have been selected for consideration here; they are arranged in order of increasing taxonomic specificity: (*a*) the breeding season, (*b*) facial movements, (*c*) the serum precipitin reaction, (*d*) diurnal vision, (*e*) macular (stereoscopic) vision, (*f*) colour vision, (*g*) drinking by suction, (*h*) finger grooming, (*i*) the specificity of the anti-erythrocyte sera, (*j*) the menstrual cycle, (*k*) the "human" nursing position, (*l*) the sexual skin, (*m*) the facts of purine metabolism, and (*n*) the blood groups. I have already tried to indicate the often flimsy basis of existing views on these questions, and the limited data that are available, but as I have pointed out in the foreword, our present "non-morphological" knowledge of the Primates should in some way be integrated, if it is to provide a basis for future organized research into the relationships of the Primates. It seems that no better way can be found than to apply existing knowledge to a scheme of phylogeny and classification.

It was noted in Chapter III that the earliest geological horizon in which Primates have been found is the Palæocene. Here, represented by tarsioid forms, they appear to have been closely connected with the menotyphlous family *Plesiadapidæ*, and when these Tarsioids are traced into the lower Eocene Formations, they are found to be almost as much differentiated from the true Lemuroids which these deposits also contain, as their modern representatives are from modern Lemuroids. Apparently nothing is known of the lemurs or Tarsioids between the close of the Eocene and the late Pleistocene (Matthew, 1928; Black, 1925).

The higher Primates are believed, on slight palæontological evidence, to have been derived from the tarsioid group. They appear for the first time in Lower Oligocene formations, where they are represented by *Parapithecus*, *Propliopithecus*, *Mœripithecus* and *Apidium*. The first named, known by a mandible, is the most primitive member of the Pithecoidea yet found, and in body-size, if its mandible is any criterion, was no bigger than the modern Talapoin monkey. Gregory (1922) describes the jaw "as tarsioid in a broad sense"; the accepted view of its teeth is that apart from foreshadowing the dentition of apes they also foreshadow those of the sub-anthropoid catarrhine Primates.

Propliopithecus, also known only by a lower jaw, was bigger than *Parapithecus*, but smaller than the Upper Miocene and Lower Pliocene *Pliopithecus*. Its teeth are those of a primitive anthropoid, and it is believed to have stood near the base of the line of descent leading to the anthropoids and to man (see Gregory, 1916). The status of *Mœripithecus* is still uncertain, but *Apidium*, again represented by a part of the mandible only, is regarded as a forerunner of the *Cercopithecidæ* (see Gregory, 1922). There is thus suggestive palæontological evidence to support the views that the tarsioid and lemuroid Primates were distinct types by the beginning of the Eocene, and that the anthropoid and cercopithecoid stocks had already diverged from each other by the close of this geological epoch. The significance of functional differentiation in the Primates can now be evaluated against this background of palæontological fact and speculation.

Uninterrupted breeding potentiality characterizes apparently all Primates except the Mascarene and probably the African Lemuroidea. As was suggested in an earlier chapter, this peculiarity of the Lemuroidea may be regarded as a specialization away from the basal breeding habit of the primate stock. The divergence of this sub-order is also evidenced by the fact that no significant reaction occurs when lemur serum is mixed with anti-human serum. In respect to this test *Tarsius* is similar, but the evidence of morphology, and the suggestive evidence about its facial movements, as well as the few facts we already have regarding its reproductive mechanisms, seem to indicate that this animal is closer than the lemurs to the main line of the descent of the Primates, even though in fact it may be widely separated from the Pithecoidea, and particularly from the more highly evolved members of that sub-order.

The New World monkeys show themselves to be part of the Pithecoid stock, in so far as their visual functions seem to be the same as those of the Old World members of this group, and in the fundamental similarity of catarrhine and platyrrhine grooming habits. The increase in the intensity of the precipitin response observed in passing from the *Hapalidæ* to Man along the lines of accepted classification, however, is suggestive of an evolutionary gradation of affinity in the sub-order Pithecoidea. It may be unwise to stress unduly the quantitative aspects of these tests, but it does seem that the weak responses obtained with the blood of New World monkeys may be taken to imply that these animals are but distantly related to the pithecoid stem from which man ultimately evolved.

Such a conclusion is strongly supported by experiments on the specificity of the red blood cells, by differences in their "nursing habits", and by the fact that the menstrual cycle appears to be wanting amongst the platyrrhine monkeys, although it characterizes the entire catarrhine division of the Pithecoidea. And since this type of œstrous cycle is part of the physiology of both the anthropoid and cercopithecoid sub-divisions of the Catarrhini, it is very unlikely that these two sub-divisions had separated by the time the Platyrrhini diverged—a conclusion supported by a great deal of morphological evidence (e.g. number of teeth, disposition of nostrils). In the light of the palæontological data bearing on the question of the sub-division of the Catarrhini, it is plain that the platyrrhine stem must have diverged during some phase of the Eocene.

The sub-division of the Old World primate stock into cercopithecoid and anthropoid stems seems to be very clearly reflected in the physiological evidence. Both branches took over from their common ancestors the potentiality for developing a sexual skin, and they also show their relationship in a certain similarity of their hæmagglutinogens. Their divergence from each other is evidenced by the existence only among the apes and man of certain specific blood groups, and apparently too by differences in their metabolism of purine bodies.

These facts relating to physiological differentiation are embodied in table X.

It is plain, as this table shows, that the distribution among the Primates of characters relating to function can be taken to support the generally accepted scheme of classification which was discussed in Chapter III. On the other hand, the consideration of functional characters gives little help in deciding between the more orthodox views of primate phylogeny and those put forward by Wood-Jones (1929) and Regan (1930).

I have already noted that Wood-Jones, on the basis of morphological evidence alone, suggests that the

TABLE X

The Phyletic Differentiation of Functional Characters among the Primates

In this table all characters are regarded as discontinuous, in the sense that no account is being taken of such differential results as those obtained with precipitin tests and tests with anti-human erythrocyte sera.

	No Breeding Season.	Facial Movements.	Serum Precipitin Reaction.	Diurnal Vision.	Macular (Stereoscopic) Vision.	Colour Vision.	Drinking by Suction.	Finger Grooming.	Related Red Cells.	Menstrual Cycle.	"Human" Nursing Position.	Sexual Skin.	Allantoin Absent.	Blood Groups A or B, or both.
Man	×	×	×	×	×	×	×	×	×	×	×	—	×	×
Apes (*Pongidæ*) .	×	×	×	×	×	×	×	×	×	×	×	×	×	×
Gibbons (*Hylobatidæ*) . . .	×	×	×	×	×	?	×	×	?	×	×	—	?	×
Old World Monkeys. . . .	×	×	×	×	×	×	×	×	×	×	×	×	—	—
New World Monkeys. . . .	×	×	×	×*	×*	×	×	×	—	—	—	—	?	—
Tarsius . . .	×	×	—	—	—	?	—	—	?	—	—	—	?	?
Asiatic Lemurs .	×	—	—	—	—	?	—	—	?	—	—	—	?	?
African Lemurs .	—	—	?	—	—	?	—	—	?	—	—	—	?	?
Mascarene Lemurs	—	—	—	—	—	—	—	—	?	—	—	—	?	—

* The genus *Aotes* is an exception.

Lemuroidea should be combined with the Menotyphla and given ordinal rank distinct from monkeys, apes, and man. Although the retention of the Lemuroidea in the Primates, or their combination with the Menotyphla, are to a large extent arbitrary matters depending on definition, Wood-Jones' support for the view that the lemurs should be excluded from the order, depends on far more than mere disagreement with the diagnostic characters of a Primate as defined by St. George Mivart (1873*a*) (see p. 16). Wood-Jones' argument is that there could not have been a lemuroid phase in the evolution of the Pithecoidea, and he bases this view largely upon the fact that the tympanic bone lies within the bulla in lemurs and outside it in all other primate forms. If one wishes to derive a pithecoid tympanic from a lemuroid form, so he writes, the law of irreversibility of evolution would have to be revoked, and if it had to be revoked here, it would also have to be revoked with numerous other lemuroid specializations, before a type sufficiently generalized to have been the progenitor of tarsioid and pithecoid Primates could be visualized.

Such argument demands the assumption, for which there is but little evidence, that it can safely be stated what was possible and what was impossible in the evolution of the characters of animals whose genetics has never been studied. Confidence on these points is not inspired by those who have studied the problem of evolutionary change in its broader aspects (*vide*, for example, De Beer, 1930). Indeed, there can be few students to-day who would dogmatically assert that a number of existing, or fossil, forms which vary in an orderly way with regard to a series of specific characters necessarily form a phyletic lineage. A comparable number of forms showing no orderly progression of a few specific characters might yet form such a phyletic line.

St. George Mivart, whose examination of the primate status of the Lemuroidea is indeed a mark of wide vision, made it quite clear that his reason for including this group of animals in the same order with monkeys and apes was not that he believed the lemurs to belong to a group ancestral to the other Primates. He regarded them as Primates because in his opinion there was "no doubt that Man, Apes, and Half-Apes together constitute a group capable of convenient and very distinct Zoological definition." In emphasizing this view, he remarked with equal clarity on the difficulties brought into the study of phylogeny by the possibilities of parallel evolution, and the even greater difficulties which anatomical science meets in attempting to decide between genetic and adaptive characters. To-day these difficulties are no less than they were sixty years ago when he wrote.

This becomes only too plain when the views of Wood-Jones and Regan are considered. To defend his contention that the lemurs should be removed from the order Primates, Wood-Jones suggests that the similarities between the brains of these animals and those of other Primates should be regarded as due to convergent development. To support the view that the Platyrrhini are derived from some primitive *Lorisidæ* and the Catarrhini from primitive *Lemuridæ*, Regan would have one believe that the similarities between the brains of New World monkeys and those of Old World monkeys are also due to convergent evolution.* But the non-cerebral characters on which both these hypotheses are based have never been proved to be of greater importance "genetically" than the cerebral characters that the Lemuroidea are stated to share with the Pithecoidea. Indeed it is difficult, almost impossible, to see how such a comparison could ever be made, or how unanimity regarding the genetic or adaptive nature of some characters could ever be reached. This is clearly shown in the points in which Wood-Jones and Regan differ. Regan, unlike Wood-Jones, apparently sees no reason why, in phylogeny, a tympanic that was once within, should later appear outside the

* On this point see Elliot-Smith, 1930*b*.

bulla. On the other hand, he regards as impossible the evolutionary transformation of one type of enamel pattern into another—a point which Wood-Jones does not even trouble to discuss. In the face of such differences of opinion, there can be little doubt that the arrangement of the Primates which should be upheld is the orthodox one, on the basis that it conforms both with an arbitrarily determined definition of a Primate, and the thesis that classification is dependent on the sum of all resemblances. St. George Mivart's words may well be quoted.

> "If any two groups of animals can easily be joined together in a larger aggregation capable of distinct definition by numerous characters, easily discernible and drawn from structures important in the economy of life, then I submit such groups should be so joined, provided they do not constitute a whole inconvenient and unmanageable from the number of its subdivisions."

The question of man's kinship with the apes is somewhat similar to this question of the primate affinities of the Lemuroidea. Man's immediate phyletic relationship to the ancestors of the anthropoid group of Primates cannot be doubted, unless it be argued that he developed the same blood groups, the same serum proteins, and the same peculiarities in purine metabolism altogether independently of the other anthropoids. Such an argument would have to be maintained by those who, like Wood-Jones, suggest that man is derived from a much more primitive stock than the ancestral stock of the existing apes. This view finds few adherents, and as Wood-Jones' critics have shown, there are no sound morphological reasons for denying man's phyletic relationship with the ape stock. When characters revealed by minute anatomical and physiological analysis segregate in accordance with what appear for presumptive reasons to be the natural sub-groups of an order, very strong contrary evidence would be needed to show that this segregation is artificial and that the presumptive reasons were false. Such evidence is not yet available, and man's phyletic relationship with the *Pongidæ* cannot therefore be seriously questioned.

The mere admission of man's kinship with the apes does not, however, take the story of his descent—and of the evolution of his fellow Primates—very far beyond the beginning of the Oligocene, beyond the period during which anthropoids were diverging from cercopithecoid forms. Having carried the tale so far, it might therefore be asked if it is not possible to carry the story of man's descent farther, and to discover whether he is more closely linked with one or more of the apes than he is with the remainder. As is of course well known, debate on this question began immediately Darwinian doctrines permeated scientific discussion. Is man descended from a chimpanzee or an orang? Are man and the chimpanzee simply collateral descendants of the same common ancestor? Was this common ancestor also the ancestor of the orang? Questions such as these have been asked for the past seventy years, and yet there is no unanimity of opinion about the correct answer. Recently Osborn (1927) has advocated the view that man's line of descent began in an Oligocene neutral stock, from which the apes were also, but independently, derived —basing his support of this view mainly on the fact that fossils of distinctly human status (e.g. Piltdown man) stretch back to the close of the Pliocene, and on differences in the mental make-up of these early men and of existing apes, differences which he infers from the material evidence of implements and from bodily structure. Gregory (1927*a*) has controverted these opinions with a reiteration of the view that man has descended from the common ancestor of the chimpanzee and gorilla during some period of the Miocene or Pliocene.* Other writers (see Klaatsch, 1923; Crookshank, 1931) derive a different kind of man from each of the existing ape stocks.

With so many conflicting views in the field, all based on the same kind of evidence, we may well agree with Osborn that the final solution of the problem of man's descent will come only when we possess a continuous series of fossils leading back from modern man to his Eocene ancestors. Meanwhile it may be asked if the question of man's precise relationship with the apes can receive an answer through the methods of the comparative anatomist.

I have already referred to the widespread and general criticisms that have been levelled against views that man and the apes were derived from different parent stocks and developed their common characters altogether independently of each other. There can be little doubt that the Hominoidea, in which superfamily Simpson (1931) places the *Hylobatidæ*, the *Pongidæ*, and the *Hominidæ*, and the Cercopithecoidea,

* Gregory argues that if man and the chimpanzee diverged from each other in the Eocene, then, given that the rate of evolution is the same in both groups, modern man should differ from the modern chimpanzee as much as the horse does from the tapir, two forms which definitely began to diverge from each other during that period; and since the differences are not equivalent, therefore man and the chimpanzee must have diverged later than did the horse and the tapir, i.e. in the Miocene, and from a chimpanzee-gorilla stock. This is but one of Gregory's arguments, but it amply demonstrates the speculative basis of his conclusions. Other papers in which he elaborates these views are 1927*b* and 1930.

comprising the Old World monkeys, can be regarded as natural sub-groups of the Catarrhini. But while it may be sterile to maintain a view that the members of such a group as the Hominoidea developed their characters without ever having been in any immediate phyletic relation with each other, it may be an altogether impossible problem to trace phyletic lineages through the different members of the group. It is possible that the common ancestors of the group transmitted to their descendants certain evolutionary potentialities, and that the same potentialities were realized at different times by types which had no closer relationship with each other than is constituted by their descent from a common ancestor. Such an hypothesis would be given a strong measure of support if evidence of the independent evolution of the same characters could be found among the members of the same family. It would then be difficult to maintain with confidence a view that because man shows a greater number of structural resemblances to the chimpanzee than to the orang, he is therefore more closely related genetically to the African than to the Eastern ape. As I have already indicated in previous chapters, physiological data do seem to provide evidence of the parallel evolution of the same characters among different members of the Old World Primates. This evidence may be briefly reviewed here.

Apparently all species of Old World Primates inherited from their common ancestor the potentiality for developing a menstrual cycle, and apparently all have developed such a cycle. This kind of reproductive rhythm is thus a group character of the Catarrhini, in the same sense that a dental formula $I\frac{2}{2}\ C\frac{1}{1}\ P\frac{2}{2}\ M\frac{3}{3}$ is. The Old World Primates also inherited the potentiality for developing a sexual skin, but this potentiality, on the other hand, was not realized by all species, for the skin has a somewhat random distribution among the *Cercopithecidæ* and among the *Pongidæ*, although this distribution on the whole does not cut across generic classification on the basis of other characters. Unless one accepts the view that the sexual skin has developed independently among different forms, it will be necessary to believe that the chimpanzee is phylogenetically more closely related to the baboons, macaques and mangabeys than to the *Colobidæ* and cercopitheques, which in turn are more intimately linked in their evolution with the gibbons and orangs. On the basis of almost any criterion but the sexual skin such an hypothesis would be grotesque. In explanation of the distribution of this character it is therefore best to fall back on Osborn's well-known postulate that descendants of a common ancestor tend to evolve on similar lines, and to regard the distribution of the sexual skin as providing evidence of parallel evolution.

It might of course be argued that the sexual skin has not developed independently among different Primates, but has been lost independently. This may be another angle from which to regard the problem, but it does not in any way alter the issue. A Primate does not inherit from its parents, and thus from its ancestors, a sexual skin; it inherits the potentiality for developing one. According to the above argument, what would then be lost is a potentiality, and this is equivalent to saying that the potentiality is not developed or evolved. In any case, recessive mutation may be as powerful an evolutionary process as dominant mutation may be. Indeed, some investigators (see Hagedoorn,

1921) believe it to be the more significant process. Thus within the confines of a related group of animals the independent development and independent loss of characters will both result in an appearance of parallel evolution.

A denial of the possibility of such evolution of physiological characters in the *Hylobatidæ*, *Pongidæ* and *Hominidæ* (which together form Simpson's Hominoidea) leads to the conclusion, as table XI shows, that man is more closely related phylogenetically to the orang, and still more particularly to the gibbon, than he is to the African apes.

TABLE XI

Differentiation of Physiological Characters among the *Hylobatidæ*, *Pongidæ* and Man

	Blood Groups.		Sexual Skin.	Allantoin.	Precipitin Test.
	A.	B.			
Man	A	B	—	—	100
Gibbon . . .	A	B	—	?	circa 95
Orang . . .	A	B	—	—	42
Chimpanzee .	A	—	+	—	100
Gorilla . . .	A	—	+(?)	?	64

Students of morphology might attempt to explain the conclusion to which this table points in the following way. On morphological grounds they would suggest that the gibbons are the most primitive of the apes (see Schultz, 1927),* a view which implies that these animals have differentiated from the fundamental primate stock and from the original anthro-

* Black (1925) expresses the very exceptional opinion that gibbons are the most specialized of the anthropoid apes.

poid stock (as typified by *Propliopithecus* of the early Oligocene) far less than have the other apes. Its physiological affinity with man might therefore be taken to imply that man is also primitive, in respect to the physiological characters under consideration. By such argument the facts regarding physiological differentiation might be brought into line with views like that of Schultz—"the evolutionary courses of man and the gorilla diverged to a lesser degree than did the ascending paths of man and the other apes." But it would be unnecessary argument, since the fact that the gibbon and man share more physiological characters with each other than man does with the chimpanzee need not necessarily be taken as proof that man is more closely related phylogenetically with the gibbon than he is with the African ape. There is suggestive evidence, as was shown in the last chapter, that man and the different apes evolved their blood groups independently, and if their blood groups, then perhaps other diversely differentiated characters too. As with the sexual skin, it may be that the common ancestors of the members of the superfamily Hominoidea transmitted to the members of this group the potentiality to develop certain characters. When these characters were uniformly developed, they became the group characters which we recognize to-day. When, however, they developed in some but not in all members of the group, it becomes almost impossible to discover whether the development in the forms exhibiting them was in any sense linked, since the characters might have been quite independently developed. Thus in the light of present physiological evidence it is difficult to envisage the basis of a phyletic scheme which links man with any one or with any two of the apes.

The same difficulty is undoubtedly experienced in morphological discussion, where, however, it is often glossed over. In a very full and clear account of the skeleton and trunk of the Primates, Schultz (1930) gives an excellent picture of the difficulty. In reference to specific skeletal characters he writes:

> "each higher primate has some skeletal features which are more highly specialized than in any other primate, but there exists not one primate of which it can be said that its entire skeleton is *the* most specialized one. . . . It is found that the gibbon and siamang possess 7 features which have evolved to a greater extreme than in any other primate, that orang-utan possesses 10, chimpanzee only 1, gorilla 7, and man 14. This means, for instance, that 14 skeletal characters are farther removed from the common ancestral condition in man than in apes, but in 23 other characters one or another ape has become more highly specialized than man. Naturally, these figures can convey only an approximate idea of the general relative degrees of specialization in the various primates and do not show the direction of such specialization. In many instances these evolutionary differentiations have proceeded in opposite directions in various types of primates, so that two quite different forms may be equally far removed from the original ancestral condition."

This picture of diverse specialization is one obtained in perhaps all comparative anatomical studies of the larger sub-groups of the Primates. The only differences between such characters as those Schultz discusses and the physiological ones considered in table V are that the anatomical characters are continuous ones (e.g. variations from straight to concave with respect to the aspect of the ventral surface of the sacrum); such characters, from the nature of the subject, are the ones mostly discussed in morphology. But the explanation for this seeming chaos in their distribution might well be parallel and orthogenetic evolution within the confines of a natural sub-group. Even if the evidence for such evolution may not be altogether clear in the field of morphology, it certainly appears to be so in that of physiology.

Palæontology is altogether unable to provide the answer to the question of man's precise relationship with the apes. At the beginning of this chapter mention was made of Primates found in the Oligocene. In the next geological horizon, the Miocene, the remains found are representative of all the modern forms, with the exception of man. *Oreopithecus*, of North Italy, has a somewhat unfamiliar dentition, which, however, shows essential resemblances to the teeth of the *Cercopithecidæ*. *Mesopithecus*, of the Upper Miocene of Greece, exhibits characters said to link the macaque with the members of the *Colobidæ*. In the late Miocene Siwalik deposits are the remains of baboons, colobus-like monkeys, macaques and cercopitheques. Apart from these monkey types, Miocene deposits contain an elaborate collection of anthropoids, almost all of which show immediate resemblances to existing apes. Recent finds in Egypt (Moghara) have shown that the bigger apes had diverged from the gibbons at the beginning of this epoch. *Pliopithecus*, commonly found all over Europe, is a gibbon. The various species of *Dryopithecus* found in Europe, India and Africa, belong to the group of which the modern chimpanzee and gorilla are members, and *Paleosimia*, of the Siwalik, is a form which is regarded as being either the immediate ancestor or a collateral relation of the modern orang. As the scene shifts to the Pliocene and Pleistocene, many remains of monkeys and apes are found which, to quote Boule (1923) "in these strata, even more than in previous ages" are "closely akin to the genera and even to the species of the present day." *

But in all the network of Tertiary forms it is impossible with even a pretence of certainty to indicate a point and declare "there stood the animal to whose evolutionary adventures man owes his presence on earth to-day." Even Gregory, who bases his views of man's kinship with a chimpanzee-gorilla stock mainly on the evidence of teeth, emphasizes "the parallelism that is often developed in the dentition of animals" (1922, p. 119). It is he, too, who has referred to the enormous variation in pattern and dimensions of fossil and recent teeth of the Hominoidea (1927). And it is owing to this variation that Pilgrim, another well-known student of fossil Primates (1927), can write "the facts relating to the fossil Anthropoids . . . are obviously incomplete, since they have proved themselves susceptible of several interpretations at the hands of competent authorities." Incomplete these facts certainly are. Students, for example, still argue whether the Piltdown teeth are those of an ape or those of a man. It is indeed plain that we have still to wait before the fossil record will be able to provide the answer to a question which studies of anatomy and physiology fail to reveal. The available evidence cannot even deny the possibility of man's independent evolution from as far back as the Oligocene, and through the Miocene up to the present day.

* Full accounts of fossil Primates can be found in Gregory (1916, 1922), Gregory and Hellman (1926), Zittel (1925), Black (1925), Pilgrim (1927), Glaessner (1931).

BIBLIOGRAPHY

ACKERMANN, K. 1898. Thierbastarde *Abh. Ber. Ver. Naturk.*, Kassel, Vol. 43, pp. 3–79.

ADAMS, D. K. 1929. Experimental Studies of Adaptive Behavior in Cats. *Comp. Psych. Monographs*, Baltimore, Vol. 6, Serial No. 27.

ALLESCH, G. J. VON. 1921. Geburt und erste Lebensmonate eines Schimpansen. *Naturwissenschaften*, Berlin, Bd. 9, pp. 774–776.

——— 1931. *Zur Nichteuklidischen Struktur des phanomenalen Raumes* (*Versuche an* Lemur mongoz mongoz L.). Fischer. Jena.

ANSON, M. L., BARCROFT, J., MIRSKY, A. E., and OINUMA, S. 1924. On the Correlation between the Spectra of various Hæmoglobins and their Relative Affinities for Oxygen and Carbon Monoxide. *Proc. Roy. Soc.*, London, B, Vol. 97, pp. 61–83.

ANTHONY, R. 1916. Le Développement du Cerveau chez les Singes. *Ann. des sc. nat.* (*Zool.*), Paris, Vol. 11, pp. 1–118.

——— 1917. La Morphologie du Cerveau chez les Singes et chez l'Homme. *Rev. Anthrop.*, Paris, Vol. 6, pp. 236–250.

ARONOVITCH, G. D. 1927. Reflexes in Apes. *J. Nerv. and Ment. Dis.*, New York, Vol. 65, pp. 457–464.

AULMANN, G. 1932. Geglückte Nachzucht eines Orang-Utan im Düsseldorfer Zoo. *Zool. Gart.*, Leipzig, Bd. 5, pp. 81–90.

BANKS, E. 1929. Interbreeding among some Bornean Leaf Monkeys of the Genus *Pithecus*. *Proc. Zool. Soc.*, London, pp. 693–695.

BATESON, W. 1928. *William Bateson, F.R.S., Naturalist.* Univ. Press. Cambridge.

BAYLIS, H. A. 1923. Some considerations on the Host-distribution of Parasitic Nematodes. *J. Linn. Soc.* (*Zool.*), London, Vol. 36, pp. 13–23.

BEER, G. R. de. 1930. *Embryology and Evolution.* Clarendon Press. Oxford.

BEEVOR, C. E., and HORSLEY, V. 1890. A record of the results obtained by Electrical Excitation of the so-called Motor Cortex and Internal Capsule in an Orang-outang (*Simia satyrus*). *Phil. Trans. Roy. Soc.*, Vol. 181, B, pp. 129–158.

BERNSTEIN, F. 1925. Zusammende Betrachtungen über die erbliche Blutstrukturen des Menschen. *Zeits. f. Abst. u. Vererb.*, Leipzig, Vol. 37, pp. 236–270.

BIEBER, I., and FULTON, J. F. 1933. The Relation of Forced Grasping and Groping to the righting Reflexes. *Amer. J. Physiol.*, Baltimore. In the Press (abstract).

BINGHAM, HAROLD C. 1927. Parental Play of Chimpanzees. *J. Mamm.*, Baltimore, Vol. 8, No. 2, pp. 77–89.

——— 1932. Gorillas in a Native Habitat. *Carnegie Inst. Pub.* Washington, No. 426.

BLACK, DAVIDSON. 1925. Asia and the Dispersal of Primates. *Bull. Geol. Soc. China*, Peking, Vol. 4, pp. 133–183.

BLYTH, E. 1863. Report of the Curator, Zoological Department. Proc. Asiat. Soc. Beng. in *J. Asiat. Soc. Beng.*, Calcutta, Vol. 32, pp. 455–456.

BOLTON, J. S. 1903. The Functions of the Frontal Lobes. *Brain*, London, Vol. 26, pp. 215–241.

BOULE, M. 1923. *Fossil Men. Elements of Human Palæontology.* Oliver and Boyd. Edinburgh.

BOYDEN, A. A. 1926. The Precipitin Reaction in the Study of Animal Relationships. *Biol. Bull.*, Woods Hole, Mass., Vol. 50, pp. 73–107.

BROWN, T. GRAHAM, and SHERRINGTON, C. S. 1911. Observations on the Localization in the Motor Cortex of the Baboon (*Papio anubis*). *J. Physiol.*, London and Cambridge, Vol. 43, pp. 209–218.

CAMERON, T. W. M. 1926. The Helminth Parasites of Animals and Human Disease. *Proc. Roy. Soc. Med.*, London (Section: Comp. Med.), Vol. 20, (5), pp. 547–556.

——— 1929. The species of *Enterobius* Leach in Primates. *J. Helminth.*, London, Vol. 7, pp. 161–182.

——— 1930. The Species of *Subulura* Molin in Primates. *J. Helminth.*, London, Vol. 8, pp. 49–58.

CARTER, J. THORNTON. 1922. On the Structure of the Enamel in the Primates and some other Mammals. *Proc. Zool. Soc.*, London, pp. 599–608.

CHINA, W. E., and UVAROV, B. P. 1932. Ecology of Man's Ancestors. *Nature*, London, Vol. 130, p. 813.

CLARK, H. C. 1930. A Preliminary Report on some Parasites in the Blood of Wild Monkeys of Panama. *Amer. J. Trop. Med.*, Baltimore, Vol. 10, pp. 25–42.

——— 1931. Progress in the Survey for Blood Parasites of the Wild Monkeys of Panama. *Amer. J. Trop. Med.*, Baltimore, Vol. 11, pp. 11–20.

CLARK, H. C., and DUNN, H. L. 1931. Experimental Efforts to Transfer Monkey Malaria to Man. *Amer. J. Trop. Med.*, Baltimore, Vol. 11, pp. 1–10.

CLARK, H. C., DUNN, H. L., and BENAVIDES, J. 1931. Experimental Transmission to Man of a Relapsing Fever Spirochete in a Wild

Monkey of Panama—*Leontocebus geoffroyi* (Pucheran). *Amer. J. Trop. Med.*, Baltimore, Vol. 11, pp. 243–257.

Clark, W. E. le Gros. 1924*a*. Notes on the Living Tarsier (*Tarsius spectrum*). *Proc. Zool. Soc.*, London, pp. 217–223.

——— 1924*b*. On the Brain of the Tree-Shrew (*Tupaia minor*). *Proc. Zool. Soc.*, London, pp. 1053–1074.

——— 1924*c*. The Myology of the Tree-Shrew (*Tupaia minor*). *Proc. Zool. Soc.*, London, pp. 461–497.

——— 1925. On the Skull of *Tupaia*. *Proc. Zool. Soc.*, London, pp. 559–567.

——— 1931. The Brain of *Microcebus murinus*. *Proc. Zool. Soc.*, London, pp. 463–486.

——— 1932*a*. The Structure and Connections of the Thalamus. *Brain*, London, Vol. 55, pp. 406–470.

——— 1932*b*. A Morphological Study of the Lateral Geniculate Body. *Brit. J. Ophthal.*, London, Vol. 16, pp. 264–284.

——— 1932*c*. An Experimental Study of Thalamic Connections in the Rat. *Phil. Trans. Roy. Soc.*, London, B, Vol. 222, pp. 1–28.

Crookshank, F. G. 1931. *The Mongol in Our Midst*. Kegan Paul. London.

Crow, W. B. 1926. Phylogeny and the Natural System. *J. Genet.*, Cambridge, Vol. 17, pp. 85–155.

Cuming, H. 1838. On the Habits of some species of Mammalia from the Philippine Islands. *Proc. Zool. Soc.*, London, Part VI, pp. 67–68.

Cuvier, G. 1817. *Le Règne Animal*. Deterville. Paris.

Cuvier, G., and E. Geoffroy Saint Hilaire. 1795. Des Caractères qui peuvent servis à divisè les singes. *Magasin Encyclopédique. Mammalogie.* Paris.

Darwin, C. 1859. *The Origin of Species*. Murray. London.

——— 1871. *The Descent of Man*. 1901 Edition. Murray. London.

Dobell, C. 1931. Researches on the Intestinal Protozoa of Monkeys and Man. IV. An Experimental Study of the *Histolytica*-like species of *Entamœba* living naturally in Macaques. *Parasitology*, Cambridge, Vol. 23, pp. 1–72.

Elliot, D. G. 1912. A Review of the Primates. *Monog. Series. Amer. Mus. Nat. Hist.*, New York, Vols. 1 to 3.

Elliot Smith, G. 1902. *Catalogue of the Physiological Series of Comparative Anatomy in the Museum of the Royal College of Surgeons of England*. London. Vol. 2.

——— 1908. On the form of the Brain in the Extinct Lemurs of Madagascar, with some remarks on the Affinities of the Indrisinæ. *Trans. Zool. Soc.*, London, Vol. 18, pp. 163–177.

ELLIOT SMITH, G. 1926. Vision and Evolution. *West London Med. J.*, London, Vol. 31, pp. 97–117.

——— 1927. *Essays on the Evolution of Man.* 2nd Edition. Oxford Univ. Press.

——— 1930*a*. *Human History*. Cape. London.

——— 1930*b*. The Classification of the Primates. *Nature*, London, Vol. 125, pp. 270–271.

ENGLISH, W. L. 1932. An exhibition of a hybrid Marmoset. *Proc. Zool. Soc.*, London, p. 1079.

EVANS, H. M., and SWEZY, O. 1929. The Chromosomes in Man. *Mem. Univ. Calif.*, Berkeley, Vol. 9, pp. 1–41.

EWING, H. E. 1927. A revision of the American Lice of the Genus *Pediculus*, together with a consideration of the significance of their Geographical and Host Distribution. *Proc. U.S. Nat. Mus.*, Washington, Vol. 68, Art. 19, pp. 1–30.

FISCHEL, W. 1930. Weitere Untersuchung der Ziele der tierischen Handlung. *Zs. vergl. Physiol.*, Berlin, Vol. 11, pp. 523–548.

FITZINGER, L. J. 1864. Nachrichten aus dem Zoologischen Garten in München. *Zool. Gart.*, Frankfurt, Vol. 5, p. 335.

FLEXNER, S., and LEWIS, P. A. 1910. Experimental Epidemic Poliomyelitis in Monkeys. *J. Exp. Med.*, New York, Vol. 12, pp. 227–255.

FLOWER, S. 1929. *Vertebrate List.* Vol. I, Mammals. Zool. Soc., London.

——— 1931. Contributions to our Knowledge of the Duration of Life in vertebrate Animals. V. Mammals. *Proc. Zool. Soc.*, London, pp. 145–234.

FLOWER, W. H., and LYDEKKER, R. 1891. *An Introduction to the Study of Mammals Living and Extinct.* Black. London.

FORBES, H. O. 1894. *Handbook to the Primates.* Allen. London.

FOX, H. 1929. The Birth of Two Anthropoid Apes. *J. Mamm.*, Baltimore, Vol. 10, pp. 37–51.

FRIEDEMANN, T. E. 1926. The Starvation Ketosis of a Monkey. *Proc. Soc. Exper. Biol. and Med.*, New York, Vol. 24, pp. 223–226.

FULTON, J. F., and KELLER, A. D. 1932. *The Sign of Babinski.* A Study of the Evolution of Cortical Dominance in Primates. Thomas. Baltimore.

FULTON, J. F., and DE BARENNE, J. G. D. 1933. The Representation of the Tail in the Motor Cortex of Primates, with special reference to Spider Monkeys. *J. Cell. and Comp. Physiol.*, Philadelphia, Vol. 2, pp. 399–426.

FULTON, J. F., JACOBSEN, C. F., and KENNARD, MARGARET A. 1932. A Note Concerning the Relation of the Frontal Lobes to Posture and Forced Grasping in Monkeys. *Brain*, London, Vol. 55, pp. 524–536.

Garrod, A. H. 1879. Notes on the Anatomy of *Gelada rueppelli*. *Proc. Zool. Soc.*, London, pp. 451–457.

Gates, R. Ruggles. 1925. Species and Chromosomes. *Amer. Nat.*, New York, Vol. 59, pp. 192–200.

Gentry, T. G. 1872. Note on a Hybrid Monkey. *Proc. Acad. Nat. Sci.*, Philadelphia, p. 122.

Geoffroy Saint Hilaire, E. 1812. *Tableau des Quadrumanes*. Paris.

Geoffroy Saint Hilaire, E., and Cuvier, G. 1824. *Histoire Naturelle des Mammifères*. Belin. Paris.

Geoffroy Saint Hilaire, I. 1851. *Catalogue des Primates*. Paris.

Gerhardt, U. 1906. Die Morphologie des Urogenital systems eines weiblichen Gorilla. *Zs. Naturwiss.*, Jena, N.S., Vol. 41, pp. 632–654.

Glaessner, M. F. 1931. Neue Zähne von Menschenaffen aus dem Miozän des Wiener Beckens. *Ann. Nat. Hist. Mus.*, Vienna, pp. 15–27.

Gray, J. E. 1870. *Catalogue of Monkeys, Lemurs and Fruit-eating Bats, in the Collection of the British Museum*. London.

Gregory, W. K. 1916. Studies on the Evolution of the Primates. Part 1. The Cope-Osborn Theory of Trituberculy and the Ancestral Molar Patterns of the Primates. Part 2. Phylogeny of Recent and Extinct Anthropoids with special reference to the Origin of Man. *Bull. Amer. Mus. Nat. Hist.*, New York, Vol. 35, pp. 239–355.

——— 1922. *The Origin and Evolution of Human Dentition*. Williams & Wilkins & Co. Baltimore.

——— 1927*a*. The Origin of Man from the Anthropoid Stem—When and Where? *Proc. Amer. Phil. Soc.*, Philadelphia, Vol. 66, pp. 439–463.

——— 1927*b*. How Near is the Relationship of Man to the Chimpanzee-Gorilla Stock? *Quart. Rev. Biol.*, Baltimore, Vol. 2, pp. 549–560.

——— 1930. A Critique of Professor Osborn's Theory of Human Origin. *Amer. J. Phys. Anthrop.*, Philadelphia, Vol. 14, pp. 133–164.

Gregory, W. K., and Hellman, M. 1926. The Dentition of Dryopithecus and the Origin of Man. *Anthrop. Pap. Amer. Mus. Nat. Hist.*, New York, Vol. 28, pp. 1–123.

Griffith, E., Hamilton-Smith, C., and Pidgeon, E. 1827. *The Class Mammalia arranged by the Baron Cuvier*. Whittaker. London.

Grünbaum, A. S. F. 1902. Note on the "Blood Relationship" of Man and the Anthropoid Apes. *Lancet*, London, Vol. 1, p. 143.

Guillaume, P., and Meyerson, I. 1930. Recherches sur l'usage de l'instrument chez les singes. I. Le problème du détour. *J. de Psych. Norm. et Path.*, Paris, Vol. 27, pp. 177–236.

GUILLAUME, P., and MEYERSON, I. 1931. Recherches sur l'usage de l'instrument chez les singes. II. l'intermédiaire lié a l'objet. *Ibid.*, Vol. 28, pp. 481–555.

GUNNING, J. W. B. 1909. Bastard zwischen *Macacus sinicus* und *Cercopithecus lalandei*. *Zool. Gart.*, Frankfurt, Vol. 51, p. 54.

HAAN, J. A. BIERENS DE. 1925*a*. Versuche über den Farbensinn der Affen. *Tydschr. Ned. Dierk. Ver.*, Leiden, Vol. 2, No. 19, pp. 71–74.

——— 1925*b*. Der relative Wert von Form und Farbenmerkmalen in der Wahrnehmung des Affen. *Biol. Zentralbl.*, Leipzig, Bd. 45, pp. 727–734.

——— 1925*c*. Experiments on Vision in Monkeys. 1. The Colour-sense of the Pig-tailed Macaque (*Nemestrinus nemestrinus*, L.). *J. Comp. Psychol.*, Baltimore, Vol. 5, pp. 417–455.

——— 1930. Über das suchen nach versteckten Futter bei Affen und Halbaffen. *Z. f. vergl. Physiol.*, Berlin, Bd. 11, pp. 630–655.

——— 1931*a*. Die Baukunst eines niederen Affen (*Cebus hypoleucus* Humb.). *Tydschr. der Ned., Dierk. Ver., Leiden*, 3de Serie, Deel 2, pp. 23–27.

——— 1931*b*. Werkzeuggebrauch und Werkzeugherstellung bei einem niederen Affen (*Cebus hypoleucus* Humb.). *Z. f. vergl. Physiol.*, Berlin, Bd. 13, pp. 640–695.

HAAN, J. A. BIERENS DE, and FRIMA, M. J. 1930. Versuche über den Farbensinn der Lemuren. *Z. f. vergl. Physiol.*, Berlin, Bd. 12, pp. 603–631.

HAGEDOORN, A. L., and HAGEDOORN, A. C. 1921. *The relative value of the processes causing evolution*. Nijhoff. The Hague.

HALDANE, J. B. S. 1932*a*. Eland-Ox Hybrid. *Nature*, London, Vol. 129, p. 906.

——— 1932*b*. *The Causes of Evolution*. Longmans, Green and Co. London.

HAMILTON, G. V. 1914. A Study of the Sexual Tendencies in Monkeys and Baboons. *J. Anim. Behav.*, Boston, Vol. 4, pp. 295–318.

HARLOW, H. F. 1932. Comparative Behavior of Primates. 3. Complicated Delayed Reaction Tests on Primates. *J. Comp. Psych.*, Baltimore, Vol. 14, pp. 241–252.

HARLOW, H. F., and ISRAEL, R. H. 1932. Comparative Behavior of Primates. 4. Delayed Reaction Tests on Subnormal Humans. *Ibid.*, Vol. 14, pp. 253–262.

HARLOW, H. F., UEHLING, H., and MASLOW, A. H. 1932. Comparative Behavior of Primates. 1. Delayed Reaction Tests on Primates from the Lemur to the Orang-Outan. *Ibid.*, Vol. 13, pp. 313–343.

HARRIS, H. A. 1929. A Preliminary Note on the Relation of Skeletal Ossification in the Hind-limb to the Index of Cerebral Value of

Anthony and Coupin. *J. Anat.*, Cambridge, Vol. 58, pp. 267–276.

HARRISON, L. W. 1931. In *A System of Bacteriology in Relation to Medicine*, Vol. 8, H.M. Stationery Office, London.

HARTMAN, C. G. 1931. The Breeding Season in Monkeys, with special reference to *Pithecus* (*Macacus*) *rhesus*. *J. Mamm.*, Baltimore, Vol. 12, pp. 129–142.

HARTMANN, R. 1886. Die weiblichen Geschlechtstheile der anthropoiden Affen und die Brunst der Affen in Allgemeinen. *Zs. Ethnol.*, Berlin, Vol. 18, pp. 431–433.

——— 1904. *Anthropoid Apes*. Kegan Paul. London.

HEGNER, R. 1928. The Evolutionary Significance of the Protozoan Parasites of Monkeys and Man. *Rev. Biol.*, Baltimore, Vol. 3, pp. 225–244.

HILL, J. P. 1932. Croonian Lecture. The Developmental History of the Primates. *Phil. Trans. Roy. Soc.*, London, B, Vol. 221, pp. 45–178.

HOGBEN, L. 1930. *The Nature of Living Matter.* Kegan Paul, Trench, Trubner & Co. London.

HOPWOOD, A. T. 1933. Miocene Primates from British East Africa. *Ann. Mag. Nat. Hist.*, London, Vol. 11, 10th series, pp. 96–98.

HRDLIČKA, A. 1905. Brain Weight in Vertebrates. *Smithson. Misc. Collect.*, Washington, Vol. 48, pp. 89–112.

——— 1925. Weight of the Brain and of the Internal Organs in American Monkeys; with data on brain weight in other Apes. *Amer. J. Phys. Anthrop.*, Washington, Vol. 8, pp. 201–211.

HUNTER, A., and GIVENS, M. H. 1912. The Metabolism of Endogenous and Exogenous Purines in the Monkey. *J. Biol. Chem.*, Baltimore, Vol. 13, pp. 371–388.

——— 1914. The Metabolism of Endogenous and Exogenous Purines in the Monkey. *Ibid.*, Vol. 17, pp. 37–53.

HUNTER, A., and WARD, F. W. 1920. Comparative studies of Purine Metabolism in various representative Mammals. *Proc. and Trans. Roy. Soc. Canada*, Ottawa, Vol. 13 (Sect. 5), pp. 7–11.

JACOBSEN, C. F. 1931. A Study of Cerebral Function in Learning. The Frontal Lobes. *J. Comp. Neurol.*, Philadelphia, Vol. 52, pp. 271–340.

——— 1933. The Influence of Premotor Area Lesions upon the Retention of Skilled Movements in Monkeys and Chimpanzees. *Research Pub. Assoc. Res. Nerv. Ment. Dis.*, Vol. 13. (In the press.)

JACOBSEN, C. F., JACOBSEN, MARION M., and YOSHIOKA, J. G. 1932. Development of an Infant Chimpanzee During her First Year. *Comp. Psych. Mon.*, Baltimore, Vol. 9, pp. 1–94.

JOLLY, W. A., and SIMPSON, SUTHERLAND. 1907. The Function of the Rolandic Cortex in Monkeys. *Proc. Roy. Soc. Edin.*, Edinburgh, Vol. 27, pp. 63–78.

JUNGEBLUT, C. W., and ENGLE, E. T. 1932. On the Property of Certain Normal Sera to Neutralise the Virus of Poliomyelitis. *Proc. Soc. Exper. Biol. and Med.*, New York, Vol. 29, pp. 879–883.

KAFKA, H. 1931. Beitrag zur Psychologie eines niederen Affen; Grössenunterscheidung bei *Cercocebus fuliginosus*. *Zs. f. vergl. Physiol.*, Berlin, Vol. 15, pp. 71–120.

KAUDERN, W. 1910. Studien über den Männlichen Geschlectsorgane von Insectivoren en Lemuriden. *Zool. Jahr.*, Jena, Anat. Vol. 31, pp. 1–106.

KEITH, A. 1895. The Growth of Brain in Men and Monkeys, with a short criticism of the usual method of stating Brain-ratios. *J. Anat. & Physiol.*, London, Vol. 29, pp. 282–303.

——— 1931. *New Discoveries Relating to the Antiquity of Man.* Williams and Norgate. London.

KELLOGG, V. L. 1913. Ectoparasites of the Monkeys, Apes, and Man. *Science*, N.S., New York, Vol. 38, pp. 601–602.

——— 1914. Ectoparasites of Mammals. *Amer. Nat.*, New York, Vol. 48, pp. 257–279.

KLAATSCH, H. 1923. *The Evolution and Progress of Mankind.* Stokes. New York.

KLÜVER, H. 1933. *Behavior Mechanisms in Monkeys.* University Press, Chicago.

KNOTTNERUS-MEYER, T. 1904. Nochmals "Über Säugertierbastarde." *Zool. Gart.*, Frankfurt, Vol. 45, pp. 60–63.

KNOWLES, R., and GUPTA, B. M. DAS. 1932. A Study of Monkey-Malaria, and its experimental transmission to Man. (A Preliminary Report.) *Ind. Med. Gaz.*, Calcutta, Vol. 67, pp. 213–268.

KÖHLER, W. 1927. *The Mentality of Apes.* Translated from the Second Revised Edition by Ella Winter, B.Sc. Kegan Paul, Trench, Trubner and Co., Ltd. London.

KOHTS, N. 1923. *Untersuchungen über die Erkentnisfähigkeiten des Schimpansen.* Moscow.

KOLMER, W. 1930. Zur Kenntnis des Auges der Primaten. *Zs. Anat. Entw.-Gesch.*, Berlin, Vol. 93, pp. 679–772.

KUHL, H. 1820. *Beiträge zur Zoologie und vergleichenden Anatomie.* Hermannschen Buchhandlung. Frankfurt am Main.

LAMARCK, J. B. 1914. *Zoological Philosophy.* Edited by H. Elliot. Macmillan. London.

LANDSTEINER, K. 1928. Sur les propriétés serologiques du sang des anthropoides. *C. R. Soc. Biol.*, Paris, Vol. 99, pp. 658–660.

LANDSTEINER, K., and LEVINE, P. 1932. Immunization of Chimpanzees with Human Blood. *J. Immunol.*, Baltimore, Vol. 22, pp. 397–400.

LANDSTEINER, K., and MILLER, C. P. 1925*a*. Serological studies on the Blood of the Primates. I. The Differentiation of Human and

Anthropoid Bloods. *J. Exp. Med.*, New York. Vol. 42, pp. 841–852.

LANDSTEINER, K., and MILLER, C. P. 1925*b*. Serological studies on the Blood of the Primates. II. The Blood Groups in Anthropoid Apes. *Ibid.*, pp. 853–863.

——— 1925*c*. Serological studies on the Blood of the Primates. III. Distribution of serological factors related to Human isoagglutinogens in the Blood of Lower Monkeys. *Ibid.*, pp. 863–872.

LANKESTER, E. RAY. 1870. *On Comparative Longevity in Man and the Lower Animals*. Macmillan. London.

LASHLEY, K. S. 1929. *Brain Mechanisms and Intelligence*. Univ. of Chicago Press. Chicago.

——— 1933. Integrative Functions of the Cerebral Cortex. *Physiol. Rev.*, Baltimore, Vol. 13, pp. 1–42.

LATTES, L. 1932. *Individuality of the Blood in Biology and in Clinical and Forensic Medicine*. Oxford Univ. Press.

LEBOUCQ, G. 1929. Le Rapport entre le Poids et la Surface de l'Hémisphère Cérébrale chez l'Homme et les Singes. *Mem. Acad. Roy. Belg.* (*Sci.*), Brussels, 2, Tome 10, pp. 3–56.

LESSON, R. P. 1827. *Manuel de Mammalogie, ou Histoire naturelle des Mammifères*. Roret. Paris.

LEYTON, A. S. F., and SHERRINGTON, C. S. 1917. Observations on the Excitable Cortex of the Chimpanzee, Orang-utan and Gorilla. *Quart. J. Exp. Physiol.*, London, Vol. 11, pp. 135–222.

LUCAS, N. S., HUME, M., and SMITH, H. H. 1927. On the Breeding of the Common Marmoset (*Hapale jacchus*, Linn.) in Captivity when irradiated with Ultra-violet Rays. *Proc. Zool. Soc.*, London, pp. 447–451.

MARSHALL, H. T. 1901–05. Studies in Hæmolysis with Special Reference to the Properties of the Blood and Body Fluids of Human Beings. *J. Exp. Med.*, New York, Vol. 6, pp. 347–375.

MASLOW, A. H., and HARLOW, H. F. 1932. Comparative Behavior of Primates. 2. Delayed Reaction Tests on Primates at Bronx Park Zoo. *J. Comp. Psych.*, Baltimore, Vol. 14, pp. 97–107.

MATTHEW, W. D. 1915. Climate and Evolution. *Ann. N.Y. Acad. Sci.*, New York, Vol. 24, pp. 171–318.

——— 1928. The Evolution of the Mammals in the Eocene. *Proc. Zool. Soc.*, London, pp. 947–985.

METCALF, M. M. 1929. Parasites and the Aid they give in Problems of Taxonomy, Geographical Distribution, and Paleogeography. *Smithson. Misc. Collect.*, Washington, Vol. 81, No. 8, pp. 1–36.

MILNE-EDWARDS, A. 1871. Observations sur quelques points de l'embryologie des Lemuriens et sur les affinités zoologiques de ces animeaux. *Ann. des sci. nat.* (Zool.), Paris, Ser. 3, Tome 15.

MITCHELL, P. C. 1911. On Longevity and Relative Viability in Mammals and Birds; with a Note on the Theory of Longevity. *Proc. Zool. Soc.*, London, Vol. 1, pp. 425–548.

——— 1912. *The Childhood of Animals.* Heinemann. London.

——— 1932. Notes on a Cow-Eland Cross. *Proc. Zool. Soc.*, London, p. 814.

MIVART, ST. GEORGE. 1873*a*. On *Lepilemur* and *Cheirogaleus*, and on the Zoological Rank of the Lemuroidea. *Proc. Zool. Soc.*, London, pp. 484–510.

——— 1873*b*. Man and Apes. Parts I and II. *Pop. Sci. Review*, London, Vol. 12, pp. 113–137, and pp. 243–264.

MORANT, G. M. 1926. Studies of Palæolithic Man. I. The Chancelade Skull and its Relation to the Modern Eskimo Skull. *Ann. Eugenics*, London, Vol. 1, pp. 257–276.

——— 1930. Studies of Palæolithic Man. IV. A Biometric Study of the Upper Palæolithic Skulls of Europe and of their Relationships to Earlier and Later Types. *Ann. Eugenics*, London, Vol. 4, Pts. I and II, pp. 109–214.

MORGAN, T. H. 1932. *The Scientific Basis of Evolution.* Norton. New York.

NELLMANN, H., and TRENDELENBURG, W. 1926. Ein Beitrag zur Intelligenzprüfung niederer Affen. *Zs. vergl. Physiol.*, Berlin, Vol. 4, pp. 142–200.

NIEMAYER, W. 1868. Züchtungserfolge im Zoologischen Garten zur Hannover. *Zool. Gart.*, Frankfurt, Vol. 9, pp. 68–72.

NISSEN, H. W. 1931. A Field Study of the Chimpanzee. *Comp. Psych. Mon.*, Baltimore, Vol. 8, pp. 1–105.

NUTTALL, G. H. F. 1904. *Blood Immunity and Blood Relationship.* Cambridge Univ. Press.

OSBORN, H. F. 1927. Recent Discoveries Relating to the Origin and Antiquity of Man. *Proc. Amer. Phil. Soc.*, Philadelphia, Vol. 66, pp. 373–389.

PAINTER, T. S. 1924. Studies in Mammalian Spermatogenesis. IV. The Sex Chromosomes in Monkeys. *J. Exp. Zool.*, Philadelphia, Vol. 39, pp. 433–451.

——— 1925. A Comparative Study of the Chromosomes of Mammals. *Amer. Nat.*, New York, Vol. 59, pp. 385–409.

——— 1928. A Comparison of the Chromosomes of the Rat and Mouse with reference to the question of Chromosome Homology in Mammals. *Genetics*, Menasha, Vol. 13, pp. 180–189.

PAVLOV, I. P. 1927. *Conditioned Reflexes.* Translated and edited by G. V. Anrep. Oxford Univ. Press.

PENROSE, L. S. 1932. The Blood Grouping of Mongolian Imbeciles. *The Lancet*, London, Feb. 20th, p. 394.

PILGRIM, G. E. 1927. A *Sivapithecus* Palate and Other Primate Fossils from India. *Palæontologica indica*, Calcutta, (N.S.) Vol. 14, pp. 1–26.

POCOCK, R. I. 1906. On the Genus *Cercocebus*, with a Key to the known Species. *Ann. Mag. Nat. Hist.*, London, Vol. 18, pp. 278–286.

——— 1907. A Monographic Revision of the Monkeys of the Genus *Cercopithecus*. *Proc. Zool. Soc.*, London, pp. 677–746.

——— 1911. Note on Hybrid Lemurs. *Proc. Zool. Soc.*, London, p. 5.

——— 1917*a*. The Lemurs of the *Hapalemur* Group. *Ann. Mag. Nat. Hist.*, London, Vol. 19, pp. 343–352.

——— 1917*b*. The Genera of *Hapalidæ* (Marmosets). *Ann. Mag. Nat. Hist.*, London, Vol. 20, pp. 247–258.

——— 1918. On the External Characters of the Lemurs and of *Tarsius*. *Proc. Zool. Soc.*, London, pp. 19–53.

——— 1920. On the External Characters of the South American Monkeys. *Proc. Zool. Soc.*, London, pp. 91–113.

——— 1921. The Systematic Value of the Glans Penis in Macaque Monkeys. *Ann. Mag. Nat. Hist.*, London, Vol. 7, pp. 224–229.

——— 1924. A New Genus of Monkeys (*Presbytiscus*). *Proc. Zool. Soc.*, London, pp. 331–332.

——— 1925*a*. Additional Notes on the External Characters of some Platyrrhine Monkeys. *Proc. Zool. Soc.*, London, pp. 27–47.

——— 1925*b*. Notes on the Cercopithecine genera *Rhinostigma* and *Miopithecus*. *Ann. Mag. Nat. Hist.*, London, Vol. 16, pp. 264–268.

——— 1925*c*. The External Characters of the Catarrhine Monkeys and Apes. *Proc. Zool. Soc.*, London, pp. 1479–1579.

——— 1927. The Gibbons of the Genus *Hylobates*. *Proc. Zool. Soc.*, London, pp. 719–741.

——— 1928. The Langurs, or Leaf Monkeys, of British India. *J. Bombay Nat. Hist. Soc.*, Vol. 32, pp. 472–504, and pp. 660–677.

——— 1931*a*. The Long-tailed Macaque Monkeys (*Macaca radiata* and *Macaca sinica*) of southern India and Ceylon. *J. Bombay Nat. Hist. Soc.*, Vol. 35, pp. 276–288.

——— 1931*b*. The Pig-tailed Macaques (*Macaca nemestrina*). *Ibid.*, pp. 297–311.

POCOCK, R. I., and Others. 1929. A Discussion on Monkeys and Human Disease. *Proc. Roy. Soc. Med.*, London, Vol. 22, Pt. 2, pp. 819–832.

POLIAK, S. 1932. The Main Afferent Fiber Systems of the Cerebral Cortex in Primates. *Univ. Calif. Pub. Anat.*, Berkeley, Vol. 2, pp. xiv–370.

PONDER, E. 1924. *The Erythrocyte and the Action of Simple Hæmolysins*. Oliver and Boyd. London.

PONDER, E., YEAGER, J. F., and CHARIPPER, H. A. 1928. Studies in Comparative Hæmatology. II. Primates. *Quart. J. Exp. Physiol.*, London, Vol. 19, pp. 181–195.

——— 1929. Hæmatology of the Primates. *Zoologica*, New York, Vol. 11, pp. 9–18.

POPOFF, I. 1929. Über einige Grössenverhältnisse der Affenhirne. *J. Psychol. und Neurol.*, Leipzig, Vol. 38, pp. 82–90.

POPOFF, I., and POPOFF, N. 1929. Beitrag zur Kenntnis der quantitativen Differenzen zwischen den Menschen- und Affenhirnen. *Ibid.*, Vol. 38, pp. 168–178.

PRICE-JONES, C. 1931. The Concentration of Hæmoglobin in Normal Human Blood. *J. Path. Bact.*, London, Vol. 34, pp. 779–789

PRYDE, J. 1931. *Recent Advances in Biochemistry*. Churchill. London.

PRZIBRAM, H. 1910. *Experimental-Zoologie*. Vol. III, Phylogenese. Deuticke, Liepzig.

REGAN, C. TATE. 1930. The Classification of the Primates. *Nature*, London, Vol. 125, pp. 125–126.

REICHERT, E. T., and BROWN, A. P. 1909. The Differentiation and Specificity of Corresponding Proteins and Other Vital Substances in Relation to Biological Classification and Organic Evolution. The Crystallography of Hæmoglobins. *Carnegie Inst. Pub. No. 116*, Washington.

ROBSON, G. C. 1928. *The Species Problem*. Oliver and Boyd. London.

RÖRIG, A. 1903. Ueber Säugetier-Bastarde. *Zool. Gart.*, Frankfurt, Vol. 44, pp. 286–292.

ROSE, STEWART. 1825. *Apology addressed to the Traveller's Club or Anecdotes of Monkeys*. Murray. London.

ROSE, W. C. 1923. Purine Metabolism. *Physiol. Rev.*, Baltimore, Vol. 3, pp. 544–602.

RUSSELL, E. S. 1932. Conation and Perception in Animal Learning. *Biol. Rev.*, Cambridge, Vol. 7, pp. 149–179.

SÀNYÀL, BABU RAM BRAMHA. 1893. Hybridization between *Semnopithecus phayrei* and *S. cristatus*. *Proc. Zool. Soc.*, London, p. 615.

SCHINZ, H. 1844. *Systematisches Verzeichnis aller bis jezt bekannten Säugethiere, oder Synopsis Mammalium*. Jent and Gassmann. Solothurn.

SCHMIDT, M. 1882. Fortpflanzung des schwarzen Maki, *Lemur niger*. *Zool. Gart.*, Frankfurt, Vol. 6, pp. 161–165.

SCHOEPFF, A. 1871. Nachrichten aus dem Zoologischen Garten in Dresden. *Zool. Gart.*, Frankfurt, Vol. 12, pp. 370–374.

Schultz, A. H. 1921. The Occurrence of a Sternal Gland in the Orang-Utan. *J. Mamm.*, Baltimore, Vol. 2, pp. 194–196.

——— 1927. Studies on the Growth of Gorilla and of other Higher Primates with special reference to a Fetus of Gorilla, preserved in the Carnegie Museum. *Mem. Carnegie Mus.*, Pittsburgh, Vol. 11, pp. 1–88.

——— 1930. The Skeleton of the Trunk and Limbs of Higher Primates. *Human Biology*, Baltimore, Vol. 2, pp. 303–438.

Schwarz, E. 1926. Die Meerkatzen der *Cercopithecus æthiops*-Gruppe. *Zs. f. Säugetierkunde*, Berlin, Vol. 1, pp. 28–47.

——— 1927. Erythrism in Monkeys of the Genus *Cercopithecus*. Studies of Variation in Mammals. *Ann. Mag. Nat. Hist.*, London, Vol. 19, pp. 151–155.

——— 1928*a*. Notes on the Classification of the African Monkeys in the Genus *Cercopithecus* Erxleben. *Ann. Mag. Nat. Hist.*, London, Vol. 1, pp. 649–663.

——— 1928*b*. Stadien der Artbildung. Die geographischen und biologischen Formen der Mona-Meerkatze (*Cercopithecus mona* Schreber). *Z. indukt. Abstaam.—v. Vererb. Lehre Suppl.*, Berlin, Vol. 2, pp. 1299–1319.

——— 1928*c*. The Species of the Genus *Cercocebus*, E. Geoffroy. *Ann. Mag. Nat. Hist.*, London, Vol. I, pp. 649–670.

——— 1931*a*. A Revision of the Genera and Species of Madagascar *Lemuridæ*. *Proc. Zool. Soc.*, London, pp. 399–428.

——— 1931*b*. On the African Long-tailed Lemurs or Galagos. *Ann. Mag. Nat. Hist.*, London, Vol. 7, pp. 41–66.

——— 1931*c*. On the African Short-tailed Lemurs or Pottos. *Ann. Mag. Nat. Hist.*, London, Vol. 8, pp. 249–256.

Sclater, P. L. 1878. Notice of Some hybrid Monkeys lately born in the Society's Menagerie. *Proc. Zool. Soc.*, London, p. 791.

——— 1885. Note on *Lemur macaco*, and the way in which it carries its Young. *Proc. Zool. Soc.*, London, pp. 672–673.

Simpson, G. G. 1931. A New Classification of Mammals. *Bull. Amer. Mus. Nat. Hist.*, New York, Vol. 59, pp. 259–293.

Snyder, L. H. 1929. *Blood Grouping in Relation to Clinical and Legal Medicine.* Ballière, Tindall and Cox. London.

Solowiev, B. M. 1930. Ueber das spezifische Gewicht des Affenblutes. *Biol. Zbl.*, Leipzig, Vol. 50, pp. 116–119.

Sonntag, C. F. 1924. *The Morphology and Evolution of the Apes and Man.* Bale, Sons and Danielsson. London.

Stiles, C. W., Hassall, A., and Nolan, M. O. 1929. Key-Catalogue of Parasites reported for Primates (Monkeys and Lemurs) with their possible Public Health importance, and Key-Catalogue of Primates for which Parasites are reported. *Hyg. Lab. Bull. No. 152*, U.S. Treasury Dept., Public Health Service, Washington.

Straus, W. L. 1929. Studies on Primate Ilia. *Am. J. Anat.*, Philadelphia, Vol. 43, pp. 403–460.

SWEZEY, W. W. 1932. The Transition of *Troglodytella abrassarti* and *Troglodytella abrassarti acuminata*, Intestinal Ciliates of the Chimpanzee, from one type to the other. *J. Parasitology*, Urbana, Ill., Vol. 19, pp. 12–16.

TEMMINCK, C. J. 1827. *Monographies de Mammalogie, ou description de quelques genres de Mammifères, dont les espèces ont étè observées dans les différens Musées de l'Europe.* Dufour and d'Ocagne. Paris.

THOMSEN, O., and KEMP. T. 1930. Blutgruppendifferenzierung bei Tieren. *Zeitschr. f. Immunitätsforsch.*, Jena, Vol. 67, pp. 251–265.

THORNDIKE, E. L. 1911. *Animal Intelligence.* Macmillan. New York.

TILNEY, F. 1928. *The Brain from Ape to Man.* Hoeben. New York. 2 Vols.

TINKLEPAUGH, O. L. 1931. Fur-picking in Monkeys as an act of Adornment. *J. Mamm.*, Baltimore, Vol. 12, p. 430.

——— 1932*a*. Parturition and Puerperal Sepsis in a Chimpanzee. *Anat. Rec.*, Philadelphia, Vol. 53, pp. 193–205.

——— 1932*b*. Multiple Delayed Reaction with Chimpanzees and Monkeys. *J. Comp. Psychol.*, Baltimore, Vol. 13, pp. 207–243.

TRENDELENBURG, W., and SCHMIDT, I. 1930. Untersuchungen über das Farbensystem der Affen. *Zs. vergl. Physiol.*, Berlin, Vol. 12, pp. 249–278.

TROISIER, J. 1928. Le groupe sanguin II de l'Homme chez le Chimpanzé. *Ann. Inst. Pasteur*, Paris, Vol. 42, pp. 363–379.

UHLENHUTH, P. 1901. Weitere Mittheilungen über meine Methode zum Nachweise von Menschenblut. *Dtsch. med. Wschr.*, Leipzig, Vol. 27, pp. 260–261.

——— 1902. Praktische Ergebnisse der forensischen Serodiagnostik des Blutes. *Dtsch. med. Wschr.*, Leipzig, Vol. 28, pp. 659–662 and pp. 679–681.

——— 1926. Die biologische Verwandtschaft zwischen Mensch und Affe. *Dtsch. med. Wschr.*, Leipzig, Vol. 52, p. 1945.

VAN HERWERDEN, M. 1905. *Bijdrage tot de Kennis van Menstrueelen Cyclus en Puerperium.* Brill. Leiden.

WATSON, J. B. 1909. Some Experiments bearing upon Color-vision in Monkeys. *J. Comp. Neurol. and Psychol.*, Philadelphia, Pa., Vol. 19, pp. 1–28.

WEINERT, H. 1932. *Ursprung der Menschheit.* Enke. Stuttgart.

WELLS, H. G. 1909–10. The Purine Metabolism of the Monkey. *J. Biol. Chem.*, Baltimore, Vol. 7, pp. 171–183.

WELLS, H. G., and CALDWELL, G. T. 1914. The Purine Enzymes of the Orang-utan (*Simia satyrus*) and Chimpanzee (*Anthropopithecus troglodytes*). *J. Biol. Chem.*, Baltimore, Vol. 18, pp. 157–165.

WENYON, C. M. 1926. *Protozoology*. Vols. 1 and 2. Ballière, Tindall and Cox. London.

WHEELER, R. H. 1929. *The Science of Psychology*. Thomas Crowell Co. New York.

WIECHOWSKI, W. 1912. Ein Beitrag zur Kenntnis des Purinstoffwechsels der Affen. *Prag. med. Woch.*, Vol. 37, pp. 275–276.

WISLOCKI, G. B., 1930. A Study of Scent Glands in the Marmosets, especially *Œdipomidas geoffroyi*. *J. Mamm.*, Baltimore, Vol. 11, pp. 475–481.

WISLOCKI, G. B., and SCHULTZ, A. H. 1925. On the Nature of Modifications of the Skin in the Sternal Region of Certain Primates. *J. Mamm.*, Baltimore, Vol. 6, pp. 236–243.

WOLFE, H. R. 1933. Factors which may modify Precipitin Tests in their Applications to Zoölogy and Medicine. *Physiol. Zoology*, Chicago, Vol. 6, pp. 55–90.

WOOD-JONES, F. 1929. *Man's Place Among the Mammals*. Arnold. London.

WOODWARD, A. S., and Others. 1920. Discussion on the Zoological Position and Affinities of *Tarsius*. *Proc. Zool. Soc.*, London, pp. 465–498.

WOOLLARD, H. H. 1925. The Anatomy of *Tarsius Spectrum*. *Proc. Zool. Soc.*, London, pp. 1071–1184.

——— 1927. The Differentiation of the Retina in Primates. *Proc. Zool. Soc.*, London, pp. 1–17.

——— 1932. In Discussion on "The Blood Groups in Genetics and Anthropology." *Brit. Med. J.*, London, ii, pp. 26–27.

YERKES, R. M. 1933. Genetic Aspects of Grooming, a Socially Important Primate Behavior Pattern. *J. Social Psych.*, Worcester, Vol. 4, pp. 3–25.

YERKES, R. M., and YERKES, A. W. 1929. *The Great Apes. A Study of Anthropoid Life*. Yale Univ. Press. New Haven.

ZITTEL, K. A. VON. 1925. *Text-book of Palæontology*. Vol. 3, Mammalia. Macmillan. London.

ZUCKERMAN, S. 1930. The Menstrual Cycle of the Primates. Part I. General Nature and Homology. *Proc. Zool. Soc.*, London, pp. 691–754.

——— 1931. The Menstrual Cycle of the Primates. Part 3. The Alleged Breeding-season of Primates, with Special Reference to the Chacma Baboon (*Papio porcarius*). *Proc. Zool. Soc.*, London, pp. 325–343.

ZUCKERMAN, S. 1932*a*. *The Social Life of Monkeys and Apes*. Kegan Paul, Trench, Trubner and Co. London.

——— 1932*b*. The Comparative Physiology of the Menstrual Cycle. *Brit. Med. J.*, London, ii, pp. 1093–1097.

——— 1932*c*. Evidence of Man's Kinship with the Primates. *Man*, London, Vol. 33, pp. 13–16.

——— 1932*d*. The Menstrual Cycle of the Primates. Part 6. Further Observations on the Breeding of Primates, with special reference to the Suborders Lemuroidea and Tarsioidea. *Proc. Zool. Soc.*, London, pp. 1059–1075.

——— 1933. Sinanthropus and Other Fossil Men: Their Relations to Each Other and to Modern Types. *Eugenics Rev.*, London, Vol. 24, pp. 273–284.

ZUCKERMAN, S., and WALLACE, H. M. 1932. The Responses of a Chacma Baboon (*P. porcarius*) in Brightness and Colour Discrimination Tests. *Proc. Zool. Soc.*, London, p. 593. (Full report in preparation.)

SUBJECT INDEX

NOTE.—Specific subjects (e.g. colour vision, blood groups) have been considered in this book in relation to as many different Primates as possible. For this reason they are not indexed for each Primate separately.

INDEX TO BIBLIOGRAPHY

For Product Safety Concerns and Information please contact our EU representative GPSR@taylorandfrancis.com Taylor & Francis Verlag GmbH, Kaufingerstraße 24, 80331 München, Germany

Batch number: 10725802

Printed by Printforce, the Netherlands